SpringerBriefs in Operations Management

Series Editor

Suresh P. Sethi
The University of Texas at Dallas, TX, USA

More information about this series at http://www.springer.com/series/13082

Salvatore Digiesi • Giuseppe Mascolo
Giorgio Mossa • Giovanni Mummolo

New Models for Sustainable Logistics

Internalization of External Costs in Inventory Management

Springer

Salvatore Digiesi
Department of Mechanics, Mathematics & Management
Polytechnic University of Bari
Bari, Italy

Giuseppe Mascolo
Department of Mechanics, Mathematics & Management
Polytechnic University of Bari
Bari, Italy

Giorgio Mossa
Department of Mechanics, Mathematics & Management
Polytechnic University of Bari
Bari, Italy

Giovanni Mummolo
Department of Mechanics, Mathematics & Management
Polytechnic University of Bari
Bari, Italy

SpringerBriefs in Operations Management
ISBN 978-3-319-19709-8 ISBN 978-3-319-19710-4 (eBook)
DOI 10.1007/978-3-319-19710-4

Library of Congress Control Number: 2015943587

Springer Cham Heidelberg New York Dordrecht London

Printed on acid-free paper

Springer International Publishing AG Switzerland is part of Springer Science+Business Media (www.springer.com)

Preface

Logistics of transport systems is a key driver for the growth of whatever economy.

Freight transport allows production systems or common citizens to receive or send materials or finished goods required by processes as well as by everyday life activities. Passenger transport, both public and private, allow people saving time for their transfers and ensure high level of mobility. Materials and people journeys fulfill economy and society expectations.

However, the overall transport sector accounts worldwide for more than half of global liquid fossil fuels consumptions which, in turn, is responsible for a nearly quarter of the world's energy-related CO_2 emissions, more than 80 % of air pollution in the cities and about 1.3 million of fatal traffic accidents per year. Negative effects represent 'external costs' paid by unaware societies and modern economies.

Costs of externalities account worldwide for more than 10 % of the GDP with an increasing trend. The European Environment Agency (EEA) and the United Nations Environment Programme (UNEP) defined the 'Avoid-Shift-Improve' (ASI) strategy to tackle the increasing of externalities while EU Commission (Directorate General for Mobility and Transport) established in 2011 a roadmap that will lead to the internalization of external costs within 2020. Research programs and strategic actions on sustainable development of smart cities are focusing on smart mobility of goods and citizens due to the relevant environmental, social and economic costs of logistics.

Internalization of cost of externalities gives rise to new logistics cost estimates and functions which managers, researchers, lecturers and students should refer in facing with logistics issues. Under this purpose the present book has been conceived.

The book focuses on freight transports of industrial production systems. The most used keywords are as follows: sustainable logistics, freight transport, internalization of external costs, environmental cost, social cost, inventory management, Economic Order Quantity—EOQ, logistics cost function, loss factor of transport, Sustainable Order Quantity—SOQ, transport means selection, stochastic variability of product demand, stochastic variability of supply lead time, sensitivity analysis, finished vehicle logistics, inland waterways, automotive supply chain, spare parts, repair policy.

The book has been subdivided into three main parts, organized as introduced below.

Chapter 1 provides a taxonomy of external cost figures as well as data set enabling the reader to perform reliable estimates of freight transport external costs. To this purpose, a full scale case study is developed.

Chapter 2 describes a new sustainable inventory management model whose cost functions include externalities. The classical 'Economic Order Quantity' model is re-formulated and the new concept of Sustainable Order Quantity (SOQ) is defined.

Finally, in Chap. 3 the SOQ model is formulated for different inventory management applications referred to both deterministic and stochastic production environments. Numerical examples are provided.

We would like to thank our colleagues, both academics and professionals from service companies, our students, and the editors at Springer for their valuable and helpful support.

Bari (Italy)

Spring 2015

Salvatore Digiesi

Giuseppe Mascolo

Giorgio Mossa

Giovanni Mummolo

Contents

List of Figures

List of Tables

About the Authors

Salvatore Digiesi. Graduated in Mechanical Engineering at the Polytechnic of Bari—Italy. European Ph.D. in "Advanced Production Systems" at the Interpolitecnica School of Doctorate. Tenured assistant professor of mechanical plants at the Polytechnic of Bari. Member of the Board of Professor of the Ph.D. Course on "Mechanical and Management Engineering" at the Polytechnic of Bari. His main research topics are sustainable production and logistics, human performance modelling, and energy recovery systems from biomasses.

Giuseppe Mascolo received his master's degree in Mechanical Engineering in 2011 at the Polytechnic of Bari and the second level master's degree in "Industrial Plant Engineering and Technologies" in 2012 at the University of Genoa, both with full marks. He is pursuing his Ph.D. in "Mechanical and Management Engineering," with a focus on Sustainable Logistics, at the Department of Mathematics, Mechanics and Management (DMMM) of the Polytechnic of Bari.

Giorgio Mossa. He earned a degree in Mechanical Engineering and a Ph.D. in "Advanced Production Systems Engineering" at the Polytechnic of Bari. He got a master in "Energy and Environmental Management and Economics" at the School "E. Mattei"—ENI Corporate University. In 2014 he earned the National Scientific Qualification for the University Associate Professor position. Tenured assistant professor of operations management and industrial systems engineering and member of the Board of the Ph.D. Course on "Mechanical and Management Engineering" at the Polytechnic of Bari. Main research topics are environmental management of production systems, human performance modelling, design and management of industrial systems, risk, safety and security management.

Giovanni Mummolo is full professor of graduate and postgraduate industrial engineering courses at the Polytechnic of Bari (Italy), Department of Mechanics, Mathematics, and Management. His main fields of research are in production management and system design. He is responsible for several international research projects and is referee of many international journals. He received the Research Award of the CIO 2014-ICIEOM-IIE International Conference. He is President of the European Academy for Industrial Management.

1 Internalization of External Costs of Freight Transport

Abstract
Accidents, global warming, congestion, air pollution and noise are examples of negative effects related to the transport activities that generate costs not fully borne by the transport users and hence not taken into account when they make a transport decision: these are the so called external costs. The internalization of the external costs of transport has been an important issue for transport research and policy development for many years worldwide. This Chapter, starting from an overview of the transport sector statistics and of transport external costs internalization in Europe, gives a taxonomy of the main transport external costs; moreover, the state of art on the cost estimation methodologies is briefly introduced. A case study from the finished vehicle logistics in the automotive sector is presented. Results show the potential external costs reduction due to the better environmental and social performance assured by the modal shift from road toward inland waterways transport.

Keywords: Sustainable logistics, Freight transport, Internalization of external costs, Environmental cost, Social cost, Finished vehicle logistics, Inland waterways

1.1 Overview on the Transport System and the Legislative Context

The transport sector, including the movement of people and goods by cars, trucks, trains, ships, airplanes, and other vehicles, is a key drive for the European Union Countries economic growth. It accounts for about the 5 % of the EU28 Gross Value Added (GVA) and employs about the 5 % of the total workforce in the EU28 [1]. In 2012, freight transport activities amounted to 3768 billion [t·km] while passenger transport ones to 6391 billion [p·km]. Figure 1.1 shows the 1995–2012 data of the Gross Domestic Product (GDP), of the freight transport and of the passenger transport in EU28 [1] (year 1995 values: 8012 billion [€] for the GDP, 3.07 billion [t·km] for the freight transport and 5.37 billion [p·km] for the passenger transport).

S. Digiesi et al., *New Models for Sustainable Logistics*,
SpringerBriefs in Operations Management, DOI 10.1007/978-3-319-19710-4_1

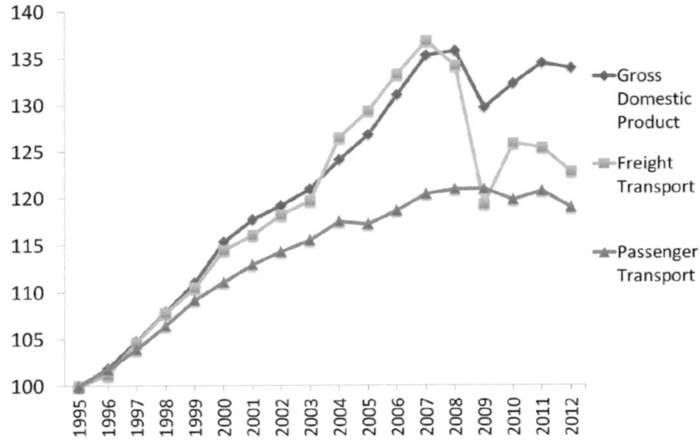

Fig. 1.1 Gross Domestic Product, passenger and freight transport trend from 1995 to 2012 in EU28 [1]

Unfortunately transport sector, characterized mainly by fossil fuel driven motor vehicles, gives rise to negative effects.

This sector, worldwide, is responsible of:

- more than half of global liquid fossil fuels consumption;
- nearly a quarter of the world's energy-related CO_2;
- more than 80 % of the air pollution in cities in developing countries;
- more than 1.27 million fatal traffic accidents per year;
- chronic traffic congestion in many of the world's urban areas.

These negative effects cause costs, which can add up to more than 10 % of a country's Gross Domestic Product, paid by the society and are likely to grow, primarily because of the expected growth of the global vehicle fleet. The continuation on a business-as-usual path will result in an increase of the global vehicle fleet from around 800 million to between 2 and 3 billion by 2050. Most of this growth will be concentrated in the developing countries. Furthermore, it is also expected an exponentially growth for the aviation sector (mainly due to the developing countries) and a growth by up to 250 % of the carbon emissions from shipping [2].

The shift toward a green transport is needed. EEA and UNEP proposed an holistic strategy, called the Avoid-Shift-Improve (ASI) strategy [2], to reach this goal. The ASI strategy aims at:

1. avoiding or reducing the number of journeys taken;
2. shifting to more environmentally efficient forms of transport;
3. improving vehicle and fuel technology.

In this context it can be included the external costs internalization strategy.

The European Commission focused the attention on the external costs of the transport for many years and, in 1995, defined the transport externalities as follows [3]:

"Transport externalities refer to a situation in which a transport user either does not pay for the full costs (e.g. including the environmental, congestion or accident costs) of his/her transport activity or does not receive the full benefits from it."

The aim of the external costs internalization is the integration of these costs in the decision making process of the transport users [4]:

- directly: through, for example, command and control measures;
- indirectly: through market-based instruments providing the right incentives to the transport users such as, for example, taxes, charges and emission trading;
- by combinations of these basic types: for example, existing taxes and charges may be differentiated by the EURO emission classes of vehicles.

The use of market-based instruments is generally regarded as the best strategy to limit the negative effects of the transport requiring, however, a detailed and reliable estimation of external costs. In order to better define the external costs it is important to highlight the difference between:

- social costs: reflect all costs occurring due to the provision and use of transport infrastructure (i.e. wear and tear costs of infrastructure, capital costs, congestion costs, accident costs, environmental costs);
- private (or internal costs): directly paid by the transport users (i.e. wear and tear and energy cost of vehicle use, own time loss costs, transport fares, and transport taxes and charges).

As aforementioned, the European Commission has pointed out the objective to charge the vehicles for the external costs they generate since 1995 [3]. European Directive 1999/62/EC [5] (also called Eurovignette-Directive) and its amendment [6] are consistent with this goal. European Directive 1999/62/EC did not includ all the transport means but was limited to vehicle taxes, tolls and user charges imposed on heavy duty vehicles (HDVs) aiming at the harmonization of levy systems and at the establishment of a fair mechanism for charging the infrastructure costs on vehicles using them. The spatial scope of this directive was the Trans-European Transport Networks (TEN-T), a planned set of road, rail, air and water transport networks to improve the transport sector performance in the European Union. The Directive already recognized, in a general way, the possibility to address a certain amount of the toll revenues to environmental protection activities but the main destination of the tolls revenues was only the recovering of the infrastructure costs (costs of construction, operation and maintenance). By adopting only the *user pays* principle, this Directive failed in recognizing also the *polluter pays* principle: all road users were considered alike without considering for example the different congestion or pollution they caused.

Furthermore, the spatial limitation to the TEN-T may cause a traffic shift towards not charged networks. In 2011, the Council adopted the new Eurovignette-Directive [6] acting on all Member States' motorways and not only on the TEN-T. Each Member State may define tolls composed of an infrastructure charge that considers also the negative effect of traffic congestion, and/or an external-cost charge related to traffic-based externalities (e.g.: air and noise pollutions). Only suggestions and not obligations are provided regarding the use of the revenue from infrastructure and external costs.

The internalization of the external cost is treated also in the EU White Paper in 2011 [7]: this document comprises 40 initiatives to be actuated within 2020 in the EU. The 'smart pricing and taxation' initiative is divided into two phases. The first phase, up to 2016, expects to start with a mandatory infrastructure charging for HDVs and to proceed with the internalization of the external costs for all modes of transport. The second phase, from 2016 until 2020, expects to implement a full and mandatory external costs internalization for road and rail transport and to examine a mandatory internalization of the external costs on all European inland waterways network. The mandatory external costs internalization could also help achieving other objectives included in the White Paper such as shifting: (i) 30 % of road freight over 300 [km] to other modes such as rail or waterborne transport by 2030 and more than 50 % by 2050; (ii) the majority of medium-distance passenger transport from road to rail by 2050.

1.2 A Taxonomy of External Costs

According to the most recent estimates, the total external costs of transport in the EU27 countries (with the exception of Malta and Cyprus but including Norway and Switzerland) in 2008 have been estimated at about 500 billion [€], excluding congestion, and at about 700 billion [€] including congestion. The GDP in EU27 in 2008 was about 12.5 quadrillion [€]: the total impact of externalities amounted to 5–6 % of GDP. Fig. 1.2a shows that accidents, congestion, climate change and air pollution represent 86 % of total costs. Moreover (see Fig. 1.2b), the road sector generate 93 % of total external costs, rail accounts for 2 %, the aviation passenger sector 4 % (only continental flights), and inland waterways 0.3 % [8].

In the following, for each cost category, the type of cost figures considered and the methodologies for their estimation are pointed out.

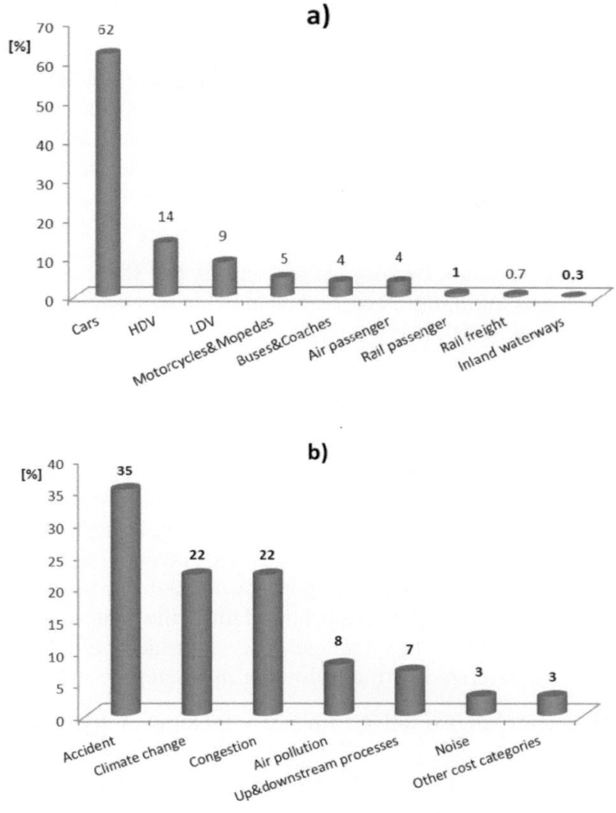

Fig. 1.2 External costs in EU27 in 2008 [8]

Accident Costs

The external accident costs are related to the costs not covered by the insurance premium such as, for example, pain and suffering caused by the traffic accidents. The best approach to estimate the marginal accident external costs is the bottom-up methodology. The main assumption of this approach is that when the driver of an additional vehicle joins the traffic exposes himself/herself to the average accident risk. This average accident risk can be estimated through the statistical relationship between the number of accidents involving a given vehicle class and the traffic flow observed in the previous years. The costs related to the accident risk are:

- the expected costs, for the person exposed to the risk, of death and injury due to an accident;
- the expected costs for the relatives and friends of the person exposed to the risk;

- accident cost for the rest of the society (material costs, police and medical costs, output costs).

The concept of the Willingness To Pay (WTP) for safety is used to evaluate the first two cost elements focusing on the Value of a Statistical Life (VSL) [4]. The estimates on the VSL generally come from studies where participants to these studies quantify own WTP for the reduction of the accident risk. These estimates are different across countries, age groups and also differ from the risk analyzed: in fact, the expected number of life years lost differs among different risk cases.

Several approaches could be adopted to quantify the share of the external costs in total accident costs taking into account what is already covered by the insurance of the person exposed to risk [4].

Climate Change Costs

Climate Change (or Global Warming) impacts of transport are mainly related to the emissions of the greenhouse gases such as carbon dioxide (CO_2), methane (CH_4) and nitrous oxide (N_2O). In the case of aviation, at high altitude, also other emissions (water vapor, sulfate, soot aerosols and nitrous oxides) have an impact on global warming. Several methodologies are available to estimate the climate change costs for the different transport modality: the state-of-art approach for evaluating this externality is the damage costs approach called Impact Pathway Approach (IPA) characterized by the following main steps:

- quantification of the GHG emission factors, in $[(CO_2)_{eq}]$ for different vehicles;
- valuation of climate change costs per tonne of $[(CO_2)_{eq}]$;
- calculation of the marginal climate change costs for different vehicle and fuel types.

The damage cost approach and the abatement cost approach are the two main methodologies evaluating the cost of the GHG emissions [4]. The first one evaluates the total costs supposing that no efforts are taken to reduce the GHG emissions; the second one evaluates the costs of achieving a certain amount of emissions reduction. Between the two methodologies the abatement cost approach is preferred, although the damage cost approach is desirable from a scientific point of view; at the same time it is characterized by a high uncertainty mainly because it is not possible to identify and to evaluate many risks related to future climate change costs. The abatement costs approach mainly reflects the willingness-to pay, of a society, for a certain abatement level of the emissions. In the abatement costs evaluation, usually two different targets are considered in Europe:

1. EU Greenhouse gases emissions reduction target for 2020 (corresponding to a cut of 20 % of GHG emissions compared to 1990 levels, “low scenario”);

2. a longer term target for keeping concentration of CO_{2eq} in the atmosphere below 450 [ppm] (thus keeping global temperature rise below 2 [°C] relative to pre-industrial levels [9], “high scenario”).

Congestion Costs

The concept of congestion externalities is easily understandable but difficult to quantify. A road network user affects, by his/her decision to use a certain network for driving between two different destinations, the utility of all other users who want to use the same network. The utility loss, aggregated over all those other users, is the negative external effect of the respective user’s decision to move between the same destinations. The utility loss is translated into costs considering the willingness to pay for avoiding this utility loss. Thus, the external effect is measured in terms of a monetary amount per trip.

The update of the unit values for congestion costs, suggested by the last “Handbook on the external costs” commissioned by the European Commission [4], is based on the aggregated approach of the FORGE model used in the National Transport Model of the United Kingdom [10].

Air Pollution Costs

Air pollution costs are mainly due to the emission of air pollutants such as particulate matter (PM_{10}, $PM_{2.5}$), nitrogen oxides (NO_x), sulfur dioxide (SO_2), ozone (O_3) and Volatile Organic Compounds (VOC). The following effects are related to this externality:

- health costs. Impacts on human health due to the aspiration of fine particles ($PM_{2.5}$/PM_{10}, other air pollutants). In addition, also Ozone (O_3) has impacts on human health;
- building and material damages. Mainly two effects have the most impact: (1) soiling of building surfaces/facades mainly through particles and dust; (2) degradation on facades and materials through corrosive processes due to acid air pollutants like NO_x and SO_2;
- crop losses in agriculture and impacts on the biosphere. Acid deposition, ozone exposition and SO_2 damage crops as well as forests and other ecosystems;
- costs for further damages for the ecosystem. Eutrophication and acidification due to the deposition of nitrogen oxides as well as contamination with heavy metals (from tire wear and tear) impact on soil and groundwater.

The unit cost estimation of the air pollution for the different transport modalities follows the already mentioned Impact Pathway Approach aiming, in this case, at the quantification of the impact of the emissions on human health (the major effects), environment, economic activity, etc. [4]. The key steps of the IPA are the:

- determination of the burden of pollutants (e.g. by using vehicle emission factors);
- modeling of the dispersion of the pollutants around the source;
- exposure assessment to evaluate the risk of the population exposed to the defined burdens;
- evaluation of the impacts caused by the pollutants to the human health and to the environment;
- monetary quantification of each impact (this step is usually based on the willingness to pay for reduced health risks).

Costs of Up and Downstream Processes

This costs category considers the external costs generated by indirect effects (not related to the transport journey itself) such as the production of energy, vehicles and transport infrastructure. These costs occur also in other markets, such as the energy market, in addition to the transport one so it is important to consider the appropriate level of internalization within these markets. The most relevant cost categories considered in [4] are:

- energy production (well-to-tank emissions—WTT);
- vehicle production, maintenance and disposal;
- infrastructure construction, maintenance and disposal.

The methodology adopted to calculate these costs is basically founded on the air pollution and climate change external costs estimation. The various studies treating this argument differ, among them, from the cost categories covered: for example some studies consider only the climate change costs of the up- and downstream processes whereas others also considers costs related to the air pollution costs.

Noise Costs

Exposure to noise emissions from traffic causes not only disturbs to people but it can affect their quality of life and health. The greater urbanization and the increase in traffic volumes are increasing the noise emissions. The two major negative impacts associated to this externality are:

- costs of annoyance: due to social disturbances of persons exposed to traffic noise which result in social and economic costs such as discomfort and pain suffering;
- health costs: noise level above 85 [dB(A)] can cause hearing damage while lower level (above 60 [dB(A)] may result in changing of heart beat frequency, increasing of blood pressure and hormonal changes. Furthermore, noise exposure can increase the risk of cardiovascular diseases and decrease the quality of sleep.

The most used methodology to estimate the marginal noise external costs is the Impact Pathway Approach. The key steps of the IPA [11] are:

- level of noise emissions measured in terms of change in time, location, frequency, level and source of noise;
- noise dispersion models used to estimate the changes in the exposure to noise according geographical locations, and measured in dB(A) and noise level indication;
- Exposure-Response Functions (ERFs) showing a relationship between decibel levels and negative impacts of the noise;
- economic valuation techniques of the negative impacts of the noise identified;
- overall assessment to identify aggregated economic values taking into account all the negative impacts identified.

Other External Costs

The researches on the external costs, generally, focus only on the most important cost categories costs (such as air pollution costs, noise costs, climate costs or accident) neglecting other external costs categories. The reasons are mainly due to the complexity in the impact pattern and uncertainty in the valuation approaches. Methodologies for the calculation of these external costs are present only in few studies thus, presently, are not as sophisticated as for the most studied external costs categories [4]. The other external costs categories estimation considered are: costs for nature and landscape, cost to ensure water and soil quality, costs to ensure biodiversity losses, cost in urban areas (such as separation costs for pedestrian and costs of scarcity for non-motorized traffic).

1.3 A Case Study from Automotive Industry Logistics

Worldwide, 65,462,496 passenger cars have been assembled and 62,786,169 registered (or sold) in 2013. Fig. 1.3 shows the worldwide passenger cars production from 2000 to 2013 [12] by macro-areas while Fig. 1.4 shows the worldwide passenger cars registrations (or sales) from 2005 to 2013 [13].

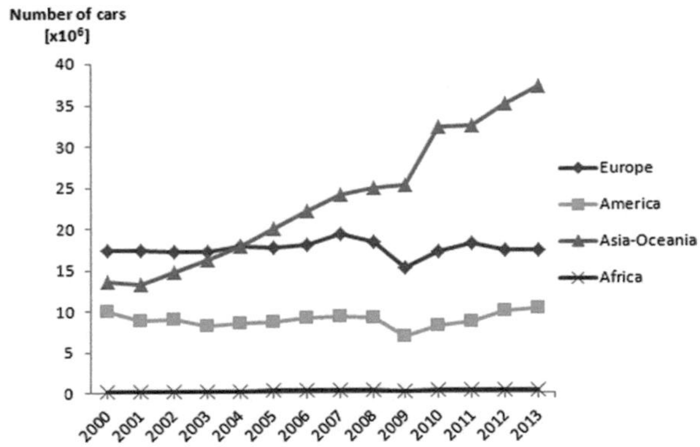

Fig. 1.3 New passenger cars assembled worldwide from 2000 to 2013 [own Figure based on [12] data]

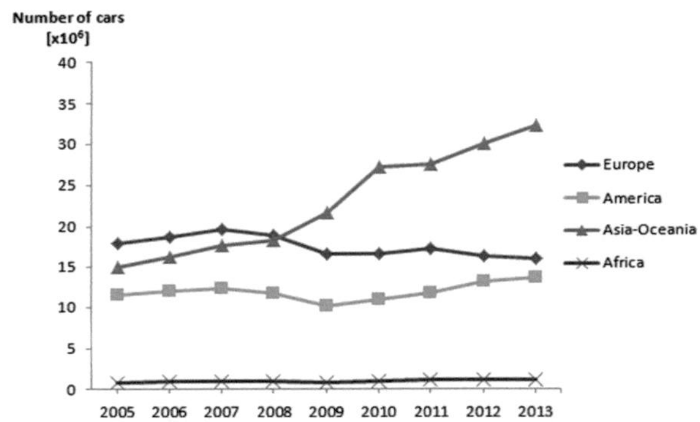

Fig. 1.4 New passenger cars registered (or sold) worldwide from 2005 to 2013 [own Figure based on [13] data]

A widespread analysis covering the Europe has been conducted to identify the flows of the new passenger cars from the assembly plants to each national European market. The main steps followed to perform the analysis have been:

- Passenger cars definition. According to Eurostat [14] and OICA [15], a passenger car is a motor vehicle with at least four wheels used for the transport of maximum nine passengers including the driver.

- Identification of automotive assembly plants in Europe. More than 100 passenger cars assembly plants have been identified in the Europe excluding Russia [16].
- Identification of new passenger car models assembled in each plant. A web-research performed on the Original Equipment Manufacturers official websites allowed identifying the car models assembled in each plant.
- Collection of new passenger cars registration statistics in 2013 for the European Countries. These statistics have been found in public databases and, for some Countries, contacting privately national statistics associations. The statistics are split by passenger cars model.
- Quantification of the new passenger cars flows from the assembly plants to the selected countries. The passenger cars distribution flows have been calculated by crossing the data collected in the previous steps. Each model assembled only in one plant (for example the Dacia Logan assembled only in the Romanian plant of Colibasi) provided quite certain information about the related distribution flows origin. In some cases, in 2013, a car model has been assembled in more than one plant (for example Audi A4 assembled in the German plants of Ingolstadt and Neckarsulm): it has been made the assumption that in each European country the 50 % of the Audi A4 flow came from Ingolstadt and 50 % from Neckarsulm. More accurate hypotheses have been made if available the total number of the cars assembled in the plants split by model. For example Opel Astra, in 2013, has been assembled in Gliwice (Poland), Rüsselsheim (Germany) and Bochum (Germany) plants, respectively 100,886, 58,547 and 16,339 units. The flows in this case have been split in proportion to the number of cars assembled in these three plants.

1.3.1 Inland Waterway Transport (IWT)

The European Union is characterized by a network of inland waterways of more than 40,000 [km] [1], 29,172 [km] of which have been earmarked by Governments as waterways of international importance [17]. The most important European waterways are located in the South-East corridor, East-West corridor, Rhine corridor and North-South corridor (Fig. 1.5). As aforementioned, inland waterway transport accounted in the EU28, in 2012, only for 4 % of freight transport based on tonne-kilometres [t·km] [1].

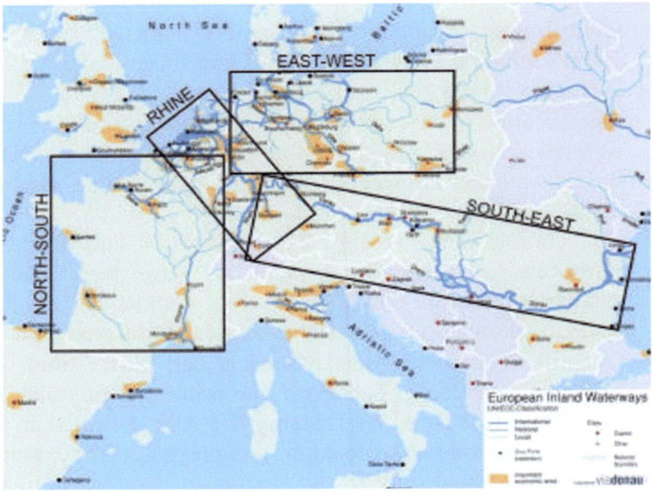

Fig. 1.5: Overview of European inland Waterways [18]

Analyses conducted for the "business as usual" outlook for 2040 [19] show that the modal share of the Inland Waterway Transport (IWT) in Europe will not increase significantly. The high capacity for bulk transport offered by IWT is presently mainly exploited by the agricultural, metal and petroleum industry and in Western Europe by the hinterland transport of maritime containers. Opportunities to raise the modal share of the IWT could be found in the following market not widely exploited by this transport modality such as:

- paper industry;
- automotive industry;
- high and heavy transport (e.g.: construction equipment, generators, turbines, wind turbines, etc.);
- waste industry.

One sector of the automotive industry with a huge potentiality for the IWT is the transport of the new passenger cars from the assembly plants to final customers. This market is, currently, not widely exploited in Europe where, however, since 1997, one of the largest European automobile logistics service providers, transports new passenger cars on the Danube waterway. New passenger cars of Ford and Renault are transported on the Danube from Kelheim to Budapest and on the way back Suzuki cars are transported from Budapest to Kelheim. Every year, Suzuki transports about 18,000 vehicles on the Danube according to a regular schedule with two departures per week by two self-propelled vessels. The "Kelheim" and the "Heilbronn" (see Fig. 1.6) can load up 200–260 cars on three decks. The loading and unloading of the vehicles are carried out through a ramp installed at the bow of the ship, which can be lowered on the Ro-Ro ramp of the port [20].

Fig. 1.6: Heilbronn vessel [21]

As aforementioned, more than 100 passenger car assembly plants were present in the EU28 in 2013 [16]: among these, 37 were present in ten Countries of the Rhine-Maine-Danube area considered in [22] (Fig. 1.7).

Fig. 1.7: Potential Countries for the distribution of the new passenger cars in the Rhine-Main-Danube area [22]

Table 1.1 shows the quantification of the potential passenger car flows among these Countries.

Table 1.1 Quantification of the potential passenger car flows in the Rhine-Main-Danube area (year 2013)

From/to	Belgium	Netherlands	Germany	Austria	Slovakia	Hungary	Serbia	Croatia	Romania	Bulgaria	Total
Belgium	16,658	18,221	71,536	8126	657	1782	165	343	595	200	118,280
Germany	123,132	89,313	1,209,523	84,476	9423	12,102	2985	6674	11,489	2952	1,552,066
Austria	2968	1824	20,286	1211	109	56	38	38	100	39	26,667
Slovakia	19,101	33,262	123,289	11,299	5559	2118	969	2046	2063	1580	201,285
Hungary	12,056	11,551	88,144	7717	3018	4380	560	628	1377	148	129,576
Serbia	5373	3246	14,512	2427	112	427	1054	34	236	57	27,476
Romania	14,855	4987	51,536	7099	2159	2932	948	672	20,908	1467	107,563

1.3.2 Discussion

A numerical example tried to quantify the potential external costs savings due to the IWT to transfer new passenger cars from four selected assembly plants located in Romania (2), Hungary (1) and Slovakia (1) to Germany, Netherlands and Belgium [23]. It has been assumed to unload: (1) in the Port of Rotterdam all the passenger cars directed to Belgium and Netherlands; (2) in the Port of Mainz the 40 % of the new cars directed to Germany. The Marco Polo Calculator has been used for calculating the external costs of the transport. The European Union's Marco Polo Programme [24] aims at shifting or avoiding freight transport off the road to other more environmentally friendly transport modes. This Programme, running by yearly calls for proposals, selects for financial support the proposals received depending on the level of the environmental and social benefits expected by them. The Marco Polo Calculator is a tool developed for the applicants of the Marco Polo Programme allowing comparing the monetized external cost impact. This calculator provides external costs estimates for road, rail, inland waterway and short sea shipping transport modes. In Table 1.2 the external costs coefficient used for the case study are shown [25].

Table 1.2 External costs of the freight transport [€cent/t·km] as in Marco Polo Calculator

Categories	Road (motorways)	Electric Train	IWT Capacity: 401–650 [t] Fuel: LNG
Air Pollution	0.858	0.100	0.180
Climate Change	0.392	0.146	0.120
Noise	0.193	0.149	–
Accidents	0.064	0.033	–
Congestion	0.343	0.020	–
Total	1.85	0.448	0.3

Vessels propelled by Liquefied Natural Gas (LNG) have been chosen to perform the case study among the many fuel technologies available for the inland waterway vessels. Table 1.3 shows an estimate of the reduction of CO_2 and air pollutants (PM, SO_2 and NO_x) emissions comparing low sulphur fuel oil and four alternative fuel technologies [25].

Table 1.4 shows the flows of the new passenger cars, in 2013, from four selected assembly plants to different national markets (Belgium, Netherlands and Germany).

Table 1.5 shows, instead, the external costs reduction of the multimodal transport (with road transport or rail transport as main haulage) compared to the direct road transport. In order to facilitate the comparison, post haulage has been neglected.

Road and rail distances, for a given pre-haulage route, have been assumed to be equal and calculated using Google Maps [26] instead Ecotransit calculation tool [27] provided the river distances for the main-haulage route.

Despite the river distance always more than 20 % longer respect to the direct road distance for all the selected routes, the multimodal transport with IWT as main haulage and rail transport as pre-haulage is the best solutions in terms of total external costs resulting in a cut of around the 80 % compared to direct road transport.

Table 1.3 CO_2 and air pollutants (PM, SO_2 and NO_x) reduction for inland vessels considering different fuel technologies

Fuel technology	NO_x	PM	SO_2	CO_2	Fuel Consumption
Fuel oil (low sulphur)			Base option		
Diesel Particulate filter (DPF)	–	−68 %	–	–	+2 %
Selective Catalytic Reduction (SCR)	−85 %	–	–	–	–
DPF + SCR	−85 %	−68 %	–	–	+2 %
Liquified Natural Gas (LNG)	−75 %	−97 %	–	−10 %	–

Table 1.4 Number of new passenger cars transported (year 2013)

OEM	Model	Average weight [t]	Assembly Plant Country (City)	Cars directed to Belgium [unit]	Cars directed to Netherlands [unit]	Cars directed to Germany [unit]
Renault	Dacia Logan	1.111	Romania	1490	65	2022
	Dacia Sandero	1.036	(Colibasi)	6274	1659	6960
	Dacia Duster	1.279		4180	151	5123
Ford	Ford B-Max	1.295	Romania (Craiova)	2911	3112	6509
Suzuki	Swift	1.055	Hungary	1849	2643	3458
	SX4	1.295	(Esztergom)	784	950	2781
	Splash	1.048		748	1338	1181
	Fiat Sedici	1.263		49	78	310
	Opel Agila	1.050		426	1569	652
Volkswagen	VW Up!	0.859	Slovakia	2095	16,187	17,147
	Skoda Citigo	0.860	(Bratislava)	685	3026	5541
	Seat Mii	0.858		379	3640	4536
	VW Touareg	2.149		181	97	3600
	Audi Q7	2.272		224	89	1332

Table 1.5 External costs reduction of multimodal transport

OEM	Assembly Plant Country (City)	Loading Port Country (City)	Pre-Haulage [km]	Unloading Port Country (City)	River Distance between Ports [km]	Road Distance Plant-Unloading Port [km]	Road transport + IWT [% reduction]	Electric train + IWT [% reduction]
Renault	Romania (Colibasi)	Romania (Calafat)	223	Germany (Mainz)	2056	1714	68	77
				Netherlands (Rotterdam)	2541	2118	70	78
Ford	Romania (Craiova)	Romania (Calafat)	91	Germany (Mainz)	2056	1637	74	78
				Netherlands (Rotterdam)	2541	2041	75	79
Suzuki	Hungary (Esztergom)	Slovakia (Sturovo)	8	Germany (Mainz)	1205	988	79	80
				Netherlands (Rotterdam)	1690	1393	80	80
Volkswagen	Slovakia (Bratislava)	Slovakia (Bratislava)	22	Germany (Mainz)	1083	862	77	79
				Netherlands (Rotterdam)	1568	1229	78	79

References

[1] EU, EU transport in figures, statistical pocketbook (2014) [Online]. Available: http://ec.europa.eu/

[2] UNEP, transport, Investing in energy and resource efficiency (2011) [Online]. Available: http://www.unep.org/

[3] EU, Towards fair and efficient pricing in transport, Policy options for internalising the external costs of transport in the European Union, Green Paper (1995) [Online]. Available: http://europa.eu/

[4] A. Korzhenevych, N. Dehnen, J. Bröcker, M. Holtkamp, M. Henning, G. Gibson, A. Varma, V. Cox, Update of the handbook on external costs of transport, Ricardo-AEA (2014)

[5] EU, European Directive 1999/62/EC: The charging of heavy goods vehicles for the use of certain infrastructures (1999) [Online]. Available: http://eur-lex.europa.eu

[6] EU, European Directive 2011/76/EU: Amending Directive 1999/62/EC on the charging of heavy goods vehicles for the use of certain infrastructures (2011) [Online]. Available: http://eur-lex.europa.eu

[7] EU, White paper on transport: roadmap to a single European transport area—towards a competitive and resource-efficient transport system (2011) [Online]. Available: http://ec.europa.eu/

[8] H. van Essen, A. Schroten, M. Otten, D. Sutter, C. Schreyer, R. Zandonella, M. Maibach, C. Doll, External costs of transport in Europe, update study (2011) [Online]. Available: http://www.cedelft.eu/

[9] M. Parry, O. Canziani, J. Palutikof, P. van der Linden, C. Hanson, *Climate Change 2007: Impacts, Adaptation and Vulnerability. IPCC Working Group II, Contribution to the Fourth Assessment Report* (Cambridge University Press, Cambridge, 2007)

[10] DfT, *National Transport Model. High Level Overview* (UK Government, Department for Transport (DfT), London, UK, 2009)

[11] S. Navrud, The state-of-the-art on economic valuation of noise—final report to European Commission DG (2002) [Online]. Available: http://ec.europa.eu/

[12] OICA, Production statistics (2014) [Online]. Available: http://www.oica.net/

[13] OICA, Sales statistics (2014) [Online]. Available: http://www.oica.net/

[14] UNECE, Illustrated glossary for transport statistics (2009) [Online]. Available: http://ec.europa.eu/

[15] OICA, Definitions [Online]. Available: http://www.oica.net/wp-content/uploads/stats-definition1.pdf. Accessed July 2014

[16] ACEA, Automobile assembly & engine production plants in Europe by country (2014) [Online]. Available: http://www.acea.be/

[17] UNECE, Inventory of main standards and parameters of the E waterway network: blue book (2012) [Online]. Available: http://www.unece.org/

[18] J. Schweighofer, The impact of extreme weather and climate change on inland waterway transport. Nat. Hazards **73**(1), 23–40 (2014)

[19] CE Delft, Planco, MDS Transmodal, viadonau and NEA, Medium and long term perspectives of Inland Waterway Transport in the European Union (2011) [Online]. Available: http://www.cedelft.eu/

[20] G. Aschauer, C. Flotzinger, C. Haide, From truck to vessel! 8 examples of modal shifts from truck to vessel (2012) [Online]. Available: www.th-wildau.de/

[21] MarineTraffic, Heilbronn [Online]. Available: http://www.marinetraffic.com/. Accessed July 2014

[22] S. Stein, H. Pascher, G. Mascolo, W. Sihn, *Potential of Using IWT for the Distribution of New Passenger Cars in Europe*. European Inland Waterway Navigation Conference 2014, Budapest, Hungary (2014)

[23] S. Digiesi, G. Mascolo, G. Mossa, G. Mummolo, *Potential Environmental Savings in New Vehicles Distribution*. Proceedings of the XIX Summer School "Francesco Turco"—2014, Senigallia, AN, Italy, 09–12 September 2014. ISBN: 978-88-908649-1-9

[24] INEA, "Marco Polo" [Online]. Available: http://inea.ec.europa.eu/. Accessed July 2014

[25] M. Brons, P. Christidis, External cost calculator for Marco Polo freight transport project proposals (2013) [Online]. Available: http://ftp.jrc.es/. Accessed July 2014

[26] Google, Google Maps [Online]. Available: https://www.google.it/maps. Accessed July 2014

[27] EcoTransIT, Calculation [Online]. Available: http://www.ecotransit.org/calculation.en.html. Accessed July 2014

2 Sustainable Inventory Management

Abstract
The Economic Order Quantity (EOQ) model, proposed by Harris in 1913, is one of the most studied models for the inventory management. The model aims at identifying the optimal lot size minimizing the total inventory costs, typically only holding and ordering costs. Many researchers extended this model trying to adapt it to real-life situation by providing new mathematical models. The increasing attention paid to sustainable manufacturing led, in the last years, to include the external costs of logistics in the EOQ model. Starting from a literature review on the inventory management models, this chapter defines the new Sustainable Order Quantity (SOQ) model. In the model, the *loss factor* parameter quantifies the loss in energy per unitary load transported and unitary distance covered, univocally identifying the various transport means. The optimal order quantity is derived minimizing a logistic cost function that considers both economic and social-environmental costs. The model allows determining at the same time the reorder level, the safety stock as well as the optimal transport means.

Keywords: Inventory management, Economic Order Quantity (EOQ), Sustainable Order Quantity (SOQ), Logistic cost function, Loss factor of transport, Environmental cost, Transport means selection

2.1 Notations

Notations adopted in this chapter are in Table 2.1.

Table 2.1 Notations adopted

Variable	Name	Unit
f	Loss factor	–
G	Product annual requirement	[unit/year]
Φ_L	Logistics costs	[€/year]
Φ_P	Purchase costs	[€/year]
Φ_O	Ordering costs	[€/year]
Φ_H	Holding costs	[€/year]
Φ_T	Transport costs	[€/year]
Φ_S	Shortage costs	[€/year]
Φ_{EX}	External costs	[€/year]
Q	Order quantity	[unit]
r	Reorder level	[unit]
H	Production hours per year	[h/year]

S. Digiesi et al., *New Models for Sustainable Logistics*,
SpringerBriefs in Operations Management, DOI 10.1007/978-3-319-19710-4_2

m	Mass of one product	[kg/unit]
D_i	Product demand of the i-th period	[unit/period]
$E(Di)$	Expected value of the product demand in the i-th period	[unit/period]
σ_{Di}	Standard deviation of the demand in the i-th period	[unit/period]
LT	Supply lead time	[h]
$E(LT)$	Expected value of the supply lead time	[h]
σ_{LT}	Standard deviation of the supply lead time	[h]
L	Transport distance	[km]
CT	Consumption time	[h]
c_H	Unit annual holding cost	[€/unit·year]
c_O	Order cost	[€/order]
SOQ	Sustainable order quantity	[unit]
SS	Safety stock level	[unit]
D^*	Maximum demand value of the i-th period not causing a stock-out event in that period	[unit/period]
D_{TOT}	Demand on LT periods	[unit/period]
D^*_{TOT}	Maximum value of the demand on LT periods not causing a stock-out event	[unit/period]
I	Inventory level	[unit]
LT^*	Maximum value of the supply lead time not causing a stock-out event at a given service level	[h]
N_S	Number of shortages in one ordering cycle	[unit]
c_S	Unitary shortage cost	[€/unit·order]
c_T	Unitary transport cost	[€/t]
(a,b,c)	Regression parameters of transport costs functions	[€/kg]
(k_1,k_2,k_3)	Regression parameters of average speed functions	[km/h]
v_{ACT}	Transport means actual speed	[km/h]
v	Transport means average speed	[km/h]
T_T	Transport time	[h]
T_L	Time required for the material handling, order management and quality control	[h]
k	Ratio between T_T and T_L	–
E_R	Energy consumption per order required by a given transport means	[MJ/order]
ε_i	Monetary cost per unit mass emission of the i-th pollutant	[€/kg]
ε_{EX}	Unit cost of externalities	[€/t·km]
E_S	Energy consumption per functional unit transported	[MJ/t·km]
$(\alpha_i, \beta_i, \gamma_i)$	Regression parameters of transport emissions functions	
(δ, μ)	Regression parameters of unit external costs functions	

2.2 Overview of the State of the Art

The Economic Order Quantity (EOQ) model [1] is one of the most investigated inventory management model aiming at identifying for a single inventory item its optimal lot size under the following hypothesis:

(i) product demand constant over time;

(ii) negligible supply lead time;

(iii) unit production cost independent from the production quantity;

(iv) zero-defective products;

(v) fixed transport costs (implicitly considered in the ordering costs).

Under these assumptions, the optimal order quantity is obtained as a trade-off between holding and order costs (Eq. 2.1).

$$EOQ = \sqrt{\frac{2 \cdot G \cdot c_O}{c_H}} \tag{2.1}$$

Starting from the work of Harris, many models have been developed including other cost figures in the optimization function.

A first class of models explicitly considers transport costs in the EOQ model. In logistic systems these costs depend on the transport means adopted, as well as on the shipment size [2], and could weight upwards of 50 % of the total logistic costs. The transport costs are explicitly considered in [3]; these costs are also considered in [4] by adopting freight rate functions available in the scientific literature. In [2] an optimal solution procedure for solving the EOQ models is provided in case of transport costs are explicitly considered and shaped as all-unit-discount costs. In [5] optimum lot-sizing algorithms in case of quantity and freight discounts are proposed: all-units and incremental discounts are considered.

In a further class of models restrictive hypotheses of [1] are released. In [6] product quality problem is investigated, and the replacing of a random quantity of defective items is considered under an EOQ model. The effects on the inventory management of imperfect quality items are considered in [7]; in the model, the percentage of the imperfect items is characterized by a known probability density function assuming also that they can be used in another production/inventory context, generating less revenue than good quality ones. The model proposed in [8] extends this model taking into account the probability of failures in inspection activities. In [9, 10] an EOQ model under the hypothesis of exponentially decaying inventory and constant product demand is defined. Authors in [11] propose a model allowing evaluating the EOQ value in case of deteriorating items and permissible supply delay.

The EOQ model proposed in [12] quantifies the increase in the optimal order quantity values due to the inflation effects on the prices; the effects of inflation uncertainty on inventory decisions are also investigated in [13]. Models including simultaneously inflation and deteriorating items effects are proposed in [14] and [15].

In [16, 17] the assumption of a deterministic demand is released, and the EOQ model is extended considering, respectively, the case of a price dependent demand and of a demand varying stochastically. Many contributions are available in which traditional EOQ model is applied under the hypothesis of stochastic variability of product demand; in case of uncertain product demand the safety stock sizing problem is integrated in the EOQ model. In [18] a model allowing obtaining optimal service level and safety stock level as a function of the ratio Q/σ_x is proposed, where σ_x is the standard deviation of the supply lead time. The evaluation of the optimal safety stock levels, through the EOQ model, of components assembled to obtain a finished product is proposed in [19]. The effects of partial backordering on EOQ solution in case of a variable product demand is evaluated in [20] by means of a deterministic inventory model: during stock out periods a fraction of the demand is backordered and the remaining fraction generates shortage costs. In [21] a solution procedure to compute EOQ in case of backordering is provided analyzing two different optimization problems, providing the optimal value of the maximum inventory level and the optimal value of the backorder level, respectively. The analysis of the potential benefits of Vendor-Managed Inventory (VMI) implementation in EOQ model is proposed in [22].

Lead time variability is investigated in the model proposed in [23]: lead time is assumed as a decision variable obtaining its optimal value by means of minimizing crashing costs, defined as extra costs to be charged in order to reduce lead time. In [24] optimal lead time value, as well as order quantity and safety stock values are obtained in case of crashing lead time costs and price discounts of backorders. In [25] an algorithm aiming at solving the single vendor single buyer problem in case of stochastic variability of lead time demand and a lead time linearly varying is proposed. The EOQ model is applied in [26] considering stochastic variability of lead time and deterministic demand rate. In [27] optimal order quantity and reorder level values are obtained in case of random lead times by assuming that it is possible to obtain expediting orders (orders with a shorter-than-average lead time at an extra cost). Methods to reduce supply lead time variability are analyzed in [28]: order splitting is identified as the optimal solution in case of a lot size-dependent supply lead time. The effect of reducing lead time variability on safety stock level, in case of a gamma distributed lead time, is investigated in [29]. Authors in [30] show how the reduction of lead time variability is more effective than the reduction of its expected value in case of a deterministic product demand.

Sustainable inventory management is a quite new research field. The increasing attention paid to sustainable manufacturing led to the development of logistic models aiming at jointly minimizing internal and external costs of logistics. Many models are available in scientific literature mainly focusing on the reduction of the carbon footprint of logistic activities. Less frequently the other categories of the external costs are considered. Authors in [31] studied the environmental impact in the inventory problem considering the carbon emissions due to the energy and fuel consumption in the transport and storage of perishable goods. In [32] authors consider an environmental performance based green cost as a linear function of the production/order quantity in the economic production quantity (EPQ) model and in the economic order quantity (EOQ) model; results of a sensitivity analysis showed that for the two models the optimal quantity value is smaller than in the standard case when green costs are accounted. In [33] social and environmental costs have been included into the classical EOQ model. Authors included social aspects in term of working hours and considered five environmental approaches to account the carbon footprint: direct accounting (carbon footprint considered as an additional economic cost), carbon tax (applied by regulatory agencies), direct cap (imposed by regulatory agencies or alternatively by public awareness about green products), cap&trade (rewards for company emitting less than an allowed cap and penalization in the other case) and carbon offsets (investments made to reduce emissions such as efficient and renewable energy resources). Similar considerations are in [34], where authors show how the overall emissions across the supply chain can be significantly reduced in case of collaborative firms within the same supply chain. In [35] authors examined the impact of the emission trading mechanism in the inventory management. They considered greenhouse gas emissions due to transport and warehousing operations. In [36] authors presented an overview about the inventory management aspects causing environmental damages that are not analyzed in depth by the traditional inventory analysis such as packaging, waste and location of stores and proposed a model that includes into the classical EOQ model environmental impacts related to vehicles journeys and waste disposal. In [37] a two-level supply chain model is proposed to determine the optimal shipment size and the optimal number of shipments taking into account the CO_2 emission costs due to transport operations. The environmental costs are split into two terms: fixed (depending on the vehicle type, vehicle age and average speed) and variable costs (depending on the actual weight of the shipment). In [38] the classical EOQ model is extended in order to solve a multi-objective problem where the goal to be achieved is not only the minimization of the logistics costs, but also the reaching of environmental and social goals. In [39] authors provided actual external costs evaluations based on a wide analysis of international researches carried out on this topic; 13 categories of external costs generated by the transport means are considered and a sensitivity analysis shows how the distribution network configuration could be influenced by a variation of these costs. In [40] authors proposed a model aiming at extending the external costs analysis of the inventory management; results of a life cycle assessment of the order quantity supplied are presented; environmental and social costs related to the transport, handling, storage, and waste disposal are considered.

In [41] authors proposed a Sustainable Order Quantity allowing evaluating the sustainable order quantity jointly minimizing the logistic and the environmental costs, as well as to identify the proper transport means under deterministic demand and lead time; the SOQ model jointly solves the two problems by the adoption of the "loss factor", a parameter which measures the loss in energy occurring during the shipping of materials by a given means of transport [42]. The global warming, the acidification and the tropospheric impacts are considered in the external costs function. In [43], authors adopted model in [41] in order to investigate the effects of different internalization strategies of external costs of transport on the optimal solution of the logistics problem. In [44] the model has been extended by considering a stochastic variability of the product demand. A further extension of the model in [41] is proposed in [45] by adding the environmental costs due to the spare parts repairing and purchasing. In [46] uncertain demand of repairable spare parts is considered. In [47] a stochastically variable lead time is considered, and the external costs due to the freight transport means as defined in European Commission commissioned studies on this issue are evaluate.

2.3 The Loss Factor of Transport

It is well known that if we want to move a load, we need to spend a certain amount of energy. This energy is partially transferred to the load, and partially lost due to the friction: the first amount could be theoretically regained but the latter have to be considered a pure loss, since there is no way to recover it.

Considering the simple case of a load W pulled over a horizontal floor (see Fig. 2.1a), it is possible to calculate the driving force F needed to move the load as:

$$F = f \cdot W, \tag{2.2}$$

with f the coefficient of friction between the lower surface of the load and the floor. In the more general case, the total amount of energy required for transport the load W along a route of length L is:

$$E = \Delta E_P + \Delta E_K + F \cdot L, \tag{2.3}$$

where

ΔE_P and ΔE_K are the differences between the initial and the final potential and kinetic energy of the load, respectively. In case of these differences are negligible, the amount of energy required is equal to $F \cdot L$. Being $F = f \cdot W$, it results

$$E = f \cdot W \cdot L. \tag{2.4}$$

When the load W is shifted for the distance L using two wheels of negligible weight (see Fig. 2.1b) a smaller amount of energy (E') is required, due to the different value assumed by the frictional coefficient: Coulomb frictional factor in the first case (*f*), a rolling frictional factor in the second one (*f'*):

$$E' = f' \cdot W \cdot L. \tag{2.5}$$

It is not possible to easily apply Eq. (2.3) in real transport system in order to compute the exact amount of energy required for the transport, but it is possible to measure the amount of energy *E* required to transport a given load *W* along a route of given length *L*.

The amount of energy required E depends on the transport means adopted. The ratio *E/(W·L)*, has been defined in [42] as the loss factor of transport (*f*). The loss factor firstly appeared in scientific literature in [42] as a technical energy-related performance measure conceived for taxonomy purposes. By adopting available data on freight transport systems, the author [42] provides a *f* taxonomy of both continuous and discontinuous transport means (see Fig. 2.2). According to the Jonkers taxonomy, a given means of transport is characterized by an *f* value, which is an average measure for a given type of transport means.

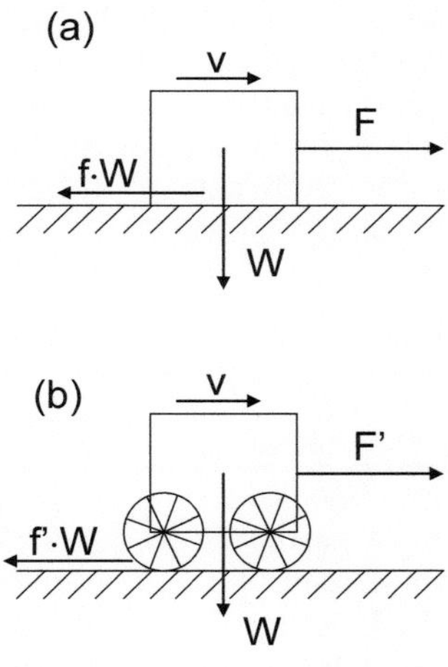

Fig. 2.1 Different ways of transporting a load

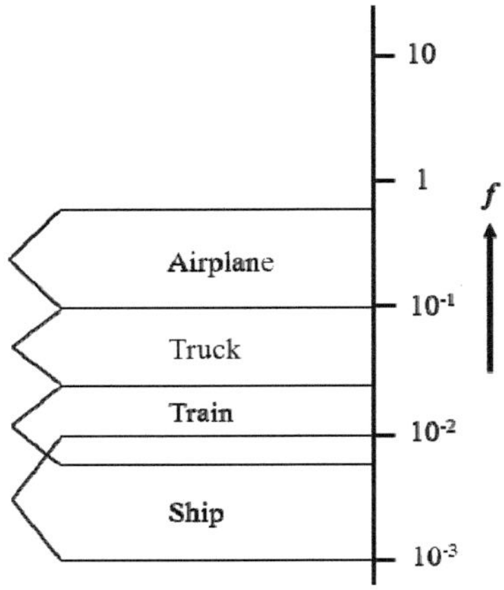

Fig. 2.2 Loss factors of different transport means (source: [42])

Loss factor values are strictly related to the energy conversion technology of transport means. In the last years, these conversion technologies significantly evolved due to technical improvements (e.g. new fuel injection technologies), as well as to policy regulations (e.g. fuel composition, emission limits). In Table 2.2 loss factor values evaluated from 2009 data of [48] database are shown. In Table 2.3 data from [48] (year 2012) and [49] (year 2012) are listed.

Table 2.2 Loss factor (*f*) value for different means of transport (data 2009)

Transportat means	*f* [J/J]
Ship—Oversee	0.010
Ship—Coast	0.033
Rail—Diesel	0.033
Rail—Electric	0.037
Ship—Inland	0.048
Truck[a] 32 ÷ 40 [t]	0.118
Truck[a] 28 ÷ 32 [t]	0.134
Truck 20 ÷ 28 [t]	0.162
Truck[a] < 28 [t]	0.170
Truck 14 ÷ 20 [t]	0.214
Truck 7.5 ÷ 14 [t]	0.242
Truck < 7.5 [t]	0.475
Small truck < 3.5 [t]	0.718
Airplane—international	0.989
Airplane—inland	1.774

[a]Operated with trailer

Table 2.3 Loss factor (f) value for different means of transport (data 2012)

Transport means	f J/J
Ship—Oversee	0.010
Ship—Tank	0.021
Ship—Coast	0.033
Rail—Diesel	0.033
Rail—Electric	0.037
Ship—Inland	0.048
Truck 34 ÷ 40 [t]	0.063
Truck 28 ÷ 34 [t]	0.069
Truck 26 ÷ 28 [t]	0.119
Truck 20 ÷ 26 [t]	0.136
Truck 14 ÷ 20 [t]	0.127
Truck 12 ÷ 14 [t]	0.139
Truck 7.5 ÷ 12 [t]	0.149
Truck 3.5 ÷ 7.5 [t]	0.165
Truck 3.5 ÷ 7.5 [t][a]	0.156
Truck < 3.5 [t][b]	0.335
Truck < 3.5 [t][c]	0.554
Airplane	0.984

[a]Truck operated without trailer

[b]Light good vehicles (LGV)

[c]Vans

2.4 A Sustainable Order Quantity (SOQ) Model

The inventory management model proposed in this work extends the definition of the Economic Order Quantity (EOQ) and defines a Sustainable Order Quantity (SOQ) as the order lot size jointly minimizing a logistics cost function Φ_L considering both logistics and external costs. By identifying with a proper loss factor f each transport means, it follows

$$\min_{SOQ,f} \Phi_L \quad [€/\text{year}]. \tag{2.6}$$

The annual logistics cost function Φ_L is defined as the sum of the purchase, ordering, holding, transport, shortage, repair, and external costs. All costs are evaluated on a yearly basis.

$$\Phi_L = \Phi_P + \Phi_O + \Phi_H + \Phi_T + \Phi_S + \Phi_{EX} \quad [€/\text{year}] \tag{2.7}$$

In the following, each cost figure of (2.7) is detailed.

2.4.1 Purchase and Ordering Costs

With G the annual product requirements, in case no quantity discount are considered, the purchase costs of the items can be computed as the product of the annual amount (G) and the unit purchase cost. In absence of quantity discount, however, purchase costs do not affect the solution of the logistics problem.

With G/Q the average number of ordering cycles in 1 year, the ordering cost is evaluated as:

$$\Phi_O = c_O \cdot \frac{G}{Q} \quad [\text{€/year}] \tag{2.8}$$

2.4.2 Transport Costs

Transport cost is the sum of the material handling and of the shipping costs. Many models are available in scientific literature allowing shaping the transport cost as a function of the transport distance for an assigned class of transport means [4, 50, 51]. For a given transport means and transport distance, the cost of transport depends on both the shipment size [2] and the speed of the transport means [52, 53]. In the model proposed, each transport means is identify by a proper loss factor value, which represents an average value of its specific energy consumption. As a consequence, for each means of transport an average characteristics speed value is considered. A single-modal transport is considered in the model. Multimodal option can be analyzed as an extension of the model proposed by considering the overall route length as the sequence of segments each operated by a different transport means. Finally, partially loaded transport means configurations are not considered.

Under these assumptions, the unit transport cost (c_T [€/t]) can be considered a function of the transport means adopted and of the transport distance:

$$c_T = c_T(f, L) \quad [\text{€/t}], \tag{2.9}$$

and the yearly transport cost can be obtained as:

$$\Phi_T = G \cdot c_T(f, L) \quad [\text{€/year}]. \tag{2.10}$$

2.4.3 Holding Costs

The holding costs are evaluated by means of different functions in the three cases considered:

(a) deterministic demand and lead time;
(b) stochastic demand and constant lead time;
(c) deterministic demand and stochastic lead time.

(a) Deterministic demand and lead time

In case of a supply lead time and a product demand constant over time (see Fig. 2.3), the holding costs can be evaluated as:

$$\Phi_H = \frac{1}{2} \cdot c_H \cdot Q \quad [€/\text{year}] \tag{2.11}$$

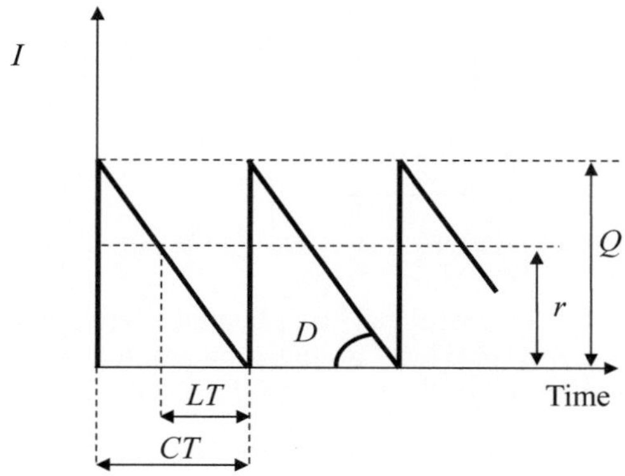

Fig. 2.3 Inventory level (I) over time in case of constant demand and lead time

(b) Stochastic demand and deterministic lead time

Stochastic variability of the product demand leads to stock-out events at a given service level (SL) (i.e. probability of no stock out). The adoption of safety stocks allows facing with stock-out events. In case of a safety stock is adopted, the holding costs are:

$$\Phi_H = c_H \cdot \left[\frac{1}{2} \cdot Q + SS \right] \quad [€/\text{year}] \tag{2.12}$$

In Fig. 2.4, order quantity Q, safety stock level SS and reorder level r are depicted.

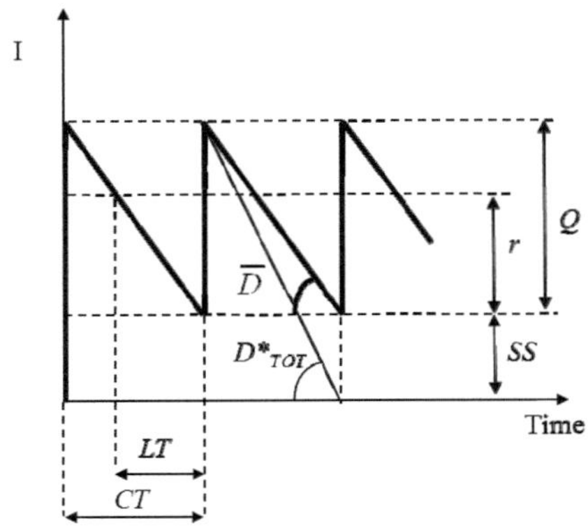

Fig. 2.4 Inventory level (*I*) over time in case of stochastic demand

The required safety stock level can be evaluated starting from the service level and the demand variability provided by the production system. With D_i the product demand of the *i*-th period of the lead time (*LT*), under the following assumptions:

- D_i are independent stochastic variables;
- product demand is characterized by the same expected value and by the same standard deviation in each *i*-th period of the lead time:

$E(D_1) = E(D_2) = \ldots = E(D_{LT}) = E(D) = G/H,$

$\sigma^2_{D_1} = \sigma^2_{D_2} = \ldots = \sigma^2_{D_{LT}} = \sigma^2_D.$

The safety stock level, *SS*, consistent with an assigned service level, *SL*, can be computed as:

$$SS = LT \cdot \left[D^* - E(D) \right] \quad [\text{unit}]. \tag{2.13}$$

The service level (SL) is:

$$SL = prob\left\{ D_{TOT} \le LT \cdot D^* \right\} = \int_{-\infty}^{LT \cdot D^*} pdf(D_{TOT})dD_{TOT}, \tag{2.14}$$

with D^* the maximum demand value of the *i*-th period not causing a stock-out in that period, D_{TOT} the demand on *LT* periods, and $LT \cdot D^* = D^*_{TOT}$ the maximum value of the demand on *LT* periods not generating a stock-out (see Fig. 2.4).

(c) <u>Deterministic demand and stochastic lead time</u>

In case of a stochastic variability of supply lead time, the expected inventory level, *E(I)*, in one ordering cycle have to be computed. The inventory level (*I*) is affected by both the order quantity size (*Q*) and the safety stock level (*SS*) as well as by the supply lead time variability.

The safety stock level value (*SS*) consistent with an assigned service level (*SL*) can be evaluated as:

$$SS = D \cdot [LT^{*} - E(LT)] \quad [\text{unit}]. \tag{2.15}$$

The service level (SL) is:

$$SL = prob\left(D_{TOT} \leq LT^{*} \cdot D\right) = \int_{-\infty}^{LT^{*}} pdf(LT)dLT = F(LT^{*}), \tag{2.16}$$

with LT^* the maximum value of the supply lead time not causing a stock out event at a given service level (*SL*) (see Fig. 2.5), and D_{TOT} the lead time demand. Introducing the standardized variable (z) of the stochastic variable *LT*, it is obtained:

$$SS = D \cdot z^{*} \cdot \sigma_{LT} \quad [\text{unit}], \tag{2.17}$$

$$SL = F(z^{*}), \tag{2.18}$$

with

$$z^{*} = \frac{LT^{*} - E(LT)}{\sigma_{LT}}. \tag{2.19}$$

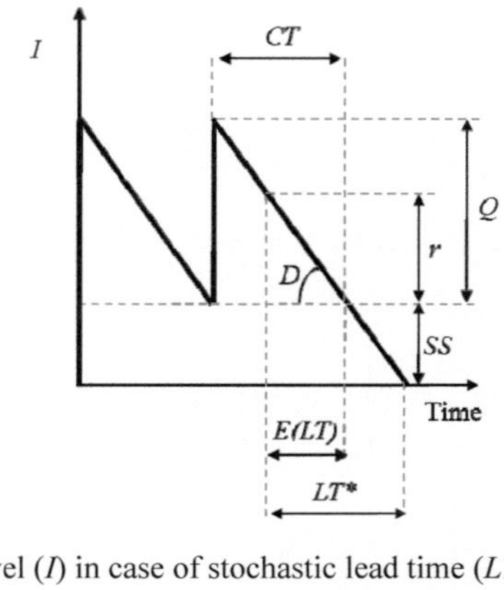

Fig. 2.5 Inventory level (*I*) in case of stochastic lead time (*LT*) and $LT = E(LT)$

Three different situations may occur due to the stochastic variability of *LT*:

(A) $LT \leq E(LT)$ (Fig. 2.6),

(B) $E(LT) < LT \leq LT^*$ (Fig. 2.7),

(C) $LT > LT^*$ (Fig. 2.8).

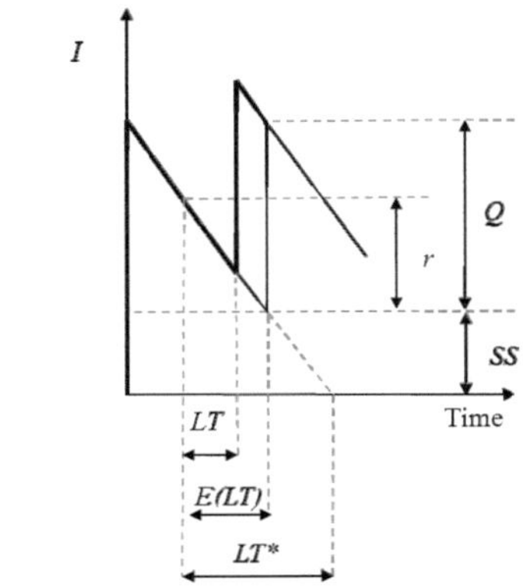

Fig. 2.6 Inventory level (I) in case of stochastic lead time (LT) and $LT \leq E(LT)$

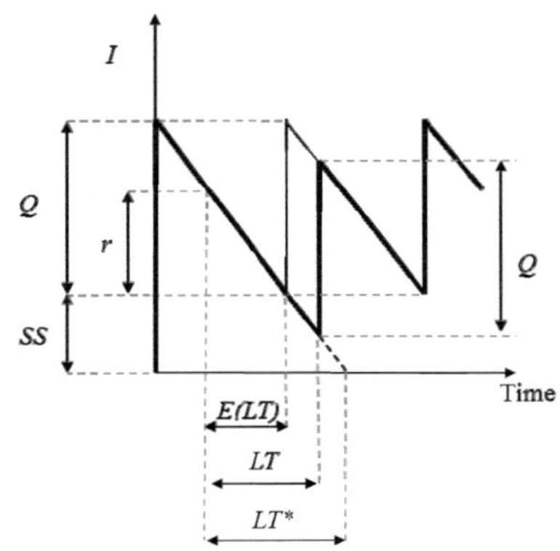

Fig. 2.7 Inventory level (I) in case of stochastic lead time (LT) and $E(LT) < LT \leq LT^*$

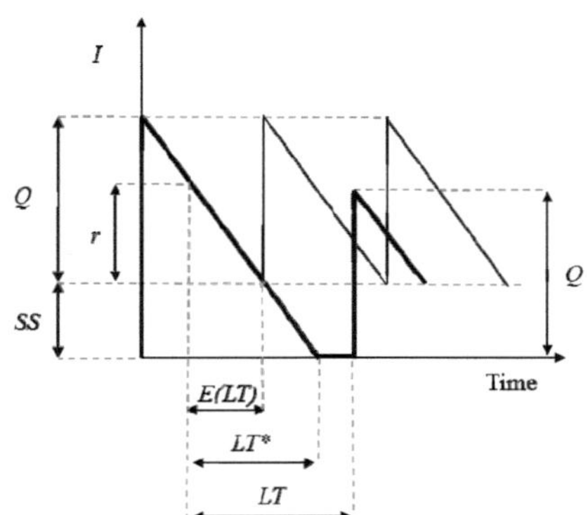

Fig. 2.8 Inventory level (I) in case of stochastic lead time (LT) and $LT > LT^*$

The expected inventory level during one ordering cycle, *E(I)*, can be evaluated as:

$$E(I) = \int_{-\infty}^{E(LT)} E(I)_A \cdot pdf(LT)dLT + \int_{E(LT)}^{LT^*} E(I)_B \cdot pdf(LT)dLT + \int_{LT^*}^{+\infty} E(I)_C \cdot pdf(LT)dLT \quad [\text{unit/order}]. \tag{2.20}$$

In the reference case of $LT = E(LT)$ (see Fig. 2.5, and the thin line in Fig. 2.6), the average inventory level and the consumption time are respectively equal to: $E(I) = 1/2 \cdot Q + SS$, and $CT = Q/D$. In cases B and C the average inventory level is affected by the lead time variability in the next ordering cycle; in case C, the length of the ordering cycle is greater than Q/D. In Table 2.4 the expected inventory level, the consumption time, and the corresponding occurrence probability values in the three cases considered are illustrated.

In case of high service level values ($SL \geq 0.90$) the following considerations arise:

- the term $[LT\text{-}LT^*]$ assumes small values;
- the occurrence probability of case C is very small;
- the consumption time in case C can be approximated to Q/D;
- the expected inventory level in case C is equal to $\frac{1}{2} \cdot Q$;
- the expected inventory level $E(I)$ in one ordering cycle is obtained as:

$$E(I) = \frac{1}{2}Q + D \cdot \sigma_{LT} \cdot [pdf(z^*) + z^* \cdot F(z^*)] \quad [\text{unit/order}], \tag{2.21}$$

and the corresponding holding cost is:

$$\Phi_H = c_H \cdot \left\{ \frac{1}{2}Q + D \cdot \sigma_{LT} \cdot [pdf(z^*) + z^* \cdot F(z^*)] \right\} \quad [€/\text{year}]. \tag{2.22}$$

Table 2.4 Expected inventory level and ordering cycle length in the three cases considered

	Expected inventory level E(I)	
Case A $LT \leq E(LT)$	$E(I_A) = \frac{1}{2}Q + D \cdot [E(LT) - LT] + SS$	
Case B $E(LT) < LT \leq LT^*$	$E(I_B) = \frac{1}{2}Q - D \cdot [LT - E(LT)] + SS$	
Case C $LT > LT^*$	$E(I_C) = \frac{1}{2}Q + \frac{1}{2}SS - \frac{1}{2}Q \cdot \frac{[LT^* - E(LT)]}{CT_C}$	
	Consumption time CT	*Case probability*
Case A $LT \leq E(LT)$	$\frac{Q}{D}$	$\int_{-\infty}^{0} pdf(z)dz$
Case B $E(LT) < LT \leq LT^*$	$\frac{Q}{D}$	$\int_{0}^{z^*} pdf(z)dz$
Case C $LT > LT^*$	$\frac{Q}{D} + [LT - LT^*]$	$\int_{z^*}^{+\infty} pdf(z)dz = 1 - SL$

2.4.4 Shortage Costs

In case of stock out events occur, shortage costs have to be considered. They can be evaluated for both the two cases:

(a) Stochastic demand and deterministic lead time
(b) Deterministic demand and stochastic lead time

(a) Stochastic demand and deterministic lead time

The number of shortage units per ordering cycle, N_S, can be calculated as:

$$N_S = \int_{D_{TOT}^*}^{+\infty} pdf(D_{TOT}) \cdot (D_{TOT} - D_{TOT}^*) \cdot dD_{TOT} \quad [\text{unit}]. \tag{2.23}$$

Under the assumptions in Sect. 2.4.3, since

$$D_{TOT} = D_1 + D_2 + \ldots + D_{LT} \quad [\text{unit/h}] \tag{2.24}$$

it follows:

$$E(D_{TOT}) = LT \cdot E(D), \tag{2.25}$$

$$\sigma_{D_{TOT}}^2 = LT \cdot \sigma_D^2. \tag{2.26}$$

For any $pdf_i(D_i)$, as stated by the central limit theorem, the $pdf(D_{TOT})$ tends towards a normal distribution:

$$pdf(D_{TOT}) = \frac{1}{\sqrt{LT} \cdot \sigma_D \cdot \sqrt{2\pi}} \cdot \exp\left[-\frac{(D_{TOT} - LT \cdot E(D))^2}{2 \cdot LT \cdot \sigma_D^2} \right]. \quad (2.27)$$

The yearly shortage cost, Φ_S, can be evaluated as:

$$\Phi_S = c_S \cdot \frac{G}{Q} \cdot \int_{D_{TOT}^*}^{+\infty} pdf(D_{TOT}) \cdot (D_{TOT} - D_{TOT}^*) dD_{TOT} \quad [\text{€/year}]. \quad (2.28)$$

(b) Deterministic demand and stochastic lead time

Stock-out events occur only in case C ($LT > LT^*$) discussed in Sect. 2.4.3, as shown in Fig. 2.8. The average number of shortage units per ordering cycle can be evaluated as:

$$N_S = D \cdot \int_{LT^*}^{+\infty} (LT - LT^*) \cdot pdf(LT) \cdot dLT \quad [\text{unit/order}], \quad (2.29)$$

and the corresponding annual shortage cost as:

$$\Phi_S = c_S \cdot \frac{G}{Q} \cdot D \cdot \sigma_{LT} \cdot L(z^*) \quad [\text{€/year}], \quad (2.30)$$

with $L(z^*)$ the standardized normal loss function defined as:

$$L(z^*) = \int_{z^*}^{+\infty} (z - z^*) \cdot pdf(z)\, dz = pdf(z^*) - z^* \cdot [1 - F(z^*)]. \quad (2.31)$$

2.4.5 External Costs

In the logistic cost function of the SOQ model (Eq. 2.7) costs of externalities are considered. As discussed in [40], different inventory management activities are responsible of environmental and social costs. The main contribution in terms of externalities has to be addressed to transport activities. Transports are responsible of environmental impacts, as well as of social damages.

As far as concern environmental impacts of transport, the related external costs can be computed by adopting two different approaches.

In [41, 44–46] the environmental costs (Φ_E [€/year]) for a given freight transport means are modelled as a function of the loss factor as:

$$\Phi_E = \frac{G}{Q}\sum_i \varepsilon_i \frac{e_i(f)}{E_S(f)} \cdot E_R \quad [€/\text{year}], \tag{2.32}$$

where:

E_S [MJ/t·km] is the energy consumption per functional unit transported;

$E_R(f) = E_S \cdot Q \cdot m \cdot L$ [MJ/order] is the energy consumption per order required by a given transport means, being m the mass of the single unit transported;

$e_i(f)$ [$kg_{i,eq}$/t·km] are the environmental impacts (i = Global Warming, GW, Acidification Potential, AP, and Tropospheric Ozone Precursor Potential, TOPP[1]) of the emissions of the transport means adopted, expressed as mass of the equivalent i-th pollutant ($i = (CO_2)_{eq.}, (SO_2)_{eq.}, NMVOC_{eq.}$) emitted per functional unit [t·km];

ε_i [€/kg] is the unit cost per impact category.

Environmental impacts are evaluated as:

$$e_i(f) = \sum_k c_{i,k} b_k(f) \quad [kg_{i,eq}/t \cdot km]. \tag{2.33}$$

where $b_k(f)$ [kg_k/t·km] is the emission factors of the k-th pollutant of a given means of transport, and $c_{i,k}$ [$kg_{i,eq}$/ kg_k] is the relative contribution (characterization factor) of the k-th burden to the i-th impact category.

An alternative approach is the one adopted in [43, 47], where estimates of external costs of freight transport available in scientific literature [54, 55], and in EU official guidelines [56, 57] are assumed.

---

[1] Tropospheric Ozone Precursor Potential; chemical compounds, such as CO, CH_4, NO_x, and Non-Methane Volatile Organic Compound (NMVOC), reacting in the presence of solar radiation and producing ozone in the troposphere; they are expressed in NMVOC equivalents, weighed factors are in [58].

References

[1] F. Harris, How many parts to make at once. Fact. Mag. Manage. **10**(2), 135–136 (1913)

[2] K. Ertogral, M. Darwish, M. Ben-Daya, Production and shipment lot sizing in a vendor–buyer supply chain with transport cost. Eur. J. Oper. Res. **176**(3), 1592–1606 (2007)

[3] W. Baumol, H. Vinod, An inventory theoretic model of freight transport demand. Manage. Sci. **16**(7), 413–421 (1970)

[4] S. Swenseth, M. Godfrey, Incorporating transport costs into inventory replenishment decisions. Int. J. Prod. Econ. **77**(2), 113–130 (2002)

[5] R. Tersine, S. Barman, Economic inventory/ transport lot sizing with quantity and freight rate discounts. Decision Sci. **22**(5), 1171–1179 (1991)

[6] X. Zhang, Y. Gerchak, Joint lot sizing and inspection policy in an EOQ model with random yield. IIE Trans. **22**(1), 41–47 (1990)

[7] M. Salameh, M. Jaber, Economic production quantity model for items with imperfect quality. Int. J. Prod. Econ. **64**, 59–64 (2000)

[8] M. Khan, M. Jaber, M. Bonney, An economic order quantity (EOQ) for items with imperfect quality and inspection errors. Int. J. Prod. Econ. **133**, 113–118 (2008)

[9] P. Ghare, C. Jaggi, A model for exponential decaying inventory. J. Ind. Eng. **14**, 238–243 (1963)

[10] R. Covert, G. Philip, An EOQ model for items with Weibull distribution deterioration. AIIE Trans. **6**, 323–326 (1973)

[11] S. Aggarwal, C. Jaggi, Ordering policies of deteriorating items under permissible delay in payment. J. Oper. Res. Soc. **6**(5), 658–662 (1995)

[12] J. Buzacott, Economic order quantities with inflation. Oper. Res. Q. **26**(3), 553–558 (1975)

[13] I. Horowitz, EOQ and inflation uncertainty. Int. J. Prod. Econ. **65**(2), 217–224 (2000)

[14] H.C. Liao, C.H. Tsai, C.T. Su, An inventory model with deteriorating items under inflation when a delay in payment is permissible. Int. J. Prod. Econ. **63**(2), 207–214 (2000)

[15] K.L. Hou, L.C. Lin, An EOQ model for deteriorating items with price- and stock-dependent selling rates under inflation and time value of money. Int. J. Syst. Sci. **37**(15), 1131–1139 (2006)

[16] T. Burwell, D. Dave, K. Fitzpatric, M. Roy, Economic lot size model for price dependent demand under quantity and freight discounts. Int. J. Prod. Econ. **48**(2), 141–155 (1997)

[17] M. Darwish, Joint determination of order quantity and reorder point of continuous review model under quantity and freight rate discounts. Comput. Oper. Res. **35**(12), 3902–3917 (2008)

[18] P. Alstrøm, Numerical computation of inventory policies, based on the EOQ/x value for order-point systems. Int. J. Prod. Econ. **71**(1–3), 235–245 (2001)

[19] M.-A. Louly, A. Dolgui, Calculating safety stocks for assembly systems with random component procurement lead times: a branch and bound algorithm. Eur. J. Oper. Res. **199**(3), 723–731 (2009)

[20] K. Park, Inventory model with partial backorders. Int. J. Syst. Sci. **13**(12), 1313–1317 (1982)

[21] R. Grubbström, A. Erdem, The EOQ with backlogging derived without derivatives. Int. J. Prod. Econ. **59**(1–3), 529–530 (1999)

[22] S. Pasandideh, S. Niaki, A. Nia, An investigation of vendor-managed inventory application in supply chain: the EOQ model with shortage. Int. J. Adv. Manuf. Technol. **49**(1–4), 329–339 (2010)

[23] M. Hariga, M. Ben-Daya, Some stochastic inventory models with deterministic variable lead time. Eur. J. Oper. Res. **113**(1), 42–51 (1999)

[24] J.-H. Pan, M.-C. Lo, Y.-C. Hsiao, Optimal reorder point inventory models with variable lead time and backorder discount considerations. Eur. J. Oper. Res. **158**(2), 488–505 (2004)

[25] M. Ben-Daya, M. Hariga, Integrated single vendor single buyer model with stochastic demand and variable lead time. Int. J. Prod. Econ. **92**(1), 75–80 (2004)

[26] M. Liberatore, The EOQ model under stochastic lead time. Oper. Res. **27**(2), 391–396 (1979)

[27] J. Bookbinder, M. Cakanyildirim, Continuous review inventory models where random lead time depends on lot size and reserved capacity. Eur. J. Oper. Res. **115**(2), 300–313 (1999)

[28] J. Hayya, D. Christy, A. Pan, Reducing inventory uncertainty: a reorder point system. Prod. Inv. Manage. J. **28**(2), 43–49 (1987)

[29] P. Wang, J. Hill, Recursive behavior of safety stock reduction: the effect of lead time uncertainty. Decision Sci. **37**(2), 285–290 (2006)

[30] X. He, J. Kim, J. Hayya, The cost of lead time variability: the case of the exponential distribution. Int. J. Prod. Econ. **97**(2), 130–142 (2005)

[31] K. Venkat, *Analyzing and Optimizing the Environmental Performance of Supply Chains*. Proceedings of the ACCEE Summer Study on Energy Efficiency in Industry (White Plains, New York, USA, 2007)

[32] Z. Tao, A. Guiffrida, M. Troutt, *A Green Cost Based Economic Production/Order Quantity Model* (Canton, OH, USA, 2010)

[33] M. Arslan, M. Turkay, *EOQ Revisited with Sustainability Considerations* (Koç University, Istanbul, Turkey, 2010)

[34] S. Benjaafar, Y. Li, M. Daskin, Carbon footprint and the management of supply chains: insights from simple models. IEEE Trans. Automat. Sci. Eng. **10**(1), 99–116 (2013)

[35] G. Hua, T. Cheng, S. Wang, Managing carbon footprints in inventory management. Int. J. Prod. Econ. **132**(2), 178–185 (2011)

[36] M. Bonney, M. Jaber, Environmentally responsible inventory models: non-classical models for a non-classical era. Int. J. Prod. Econ. **133**(1), 43–53 (2011)

[37] M. Wahab, S. Mamum, P. Ongkunaruk, EOQ models for a coordinated two-level international supply chain considering imperfect items and environmental impact. Int. J. Prod. Econ. **134**, 151–158 (2011)

[38] Y. Bouchery, A. Ghaffari, Z. Jemai, Y. Dallery, Including sustainability criteria into inventory models. Eur. J. Oper. Res. **222**(2), 229–240 (2012)

[39] C. Ortolani, A. Persona, F. Sgarbossa, External cost effects and freight modal choice: research and application. Int. J. Logist. Res. Appl. **14**(3), 199–200 (2011)

[40] D. Battini, A. Persona, F. Sgarbossa, A sustainable EOQ model: theoretical formulation and applications. Int. J. Prod. Econ. **149**(1), 145–153 (2014)

[41] S. Digiesi, G. Mossa, G. Mummolo, A loss factor based approach for sustainable logistic. Prod. Plann. Control **23**(2–3), 160–170 (2012)

[42] C. Jonkers, The loss factor of transport. Förden Heben **31**(2), 98–101 (1981)

[43] S. Digiesi, G. Mascolo, G. Mossa, G. Mummolo, Internalisation strategies of external costs of transport for a sustainable logistics. Int. J. Prod. Qual. Manage (2015)

[44] S. Digiesi, G. Mossa, G. Mummolo, *A Sustainable Order Quantity Model Under Uncertain Product Demand*. Manufacturing Modelling, Management, and Control, IFAC Proceedings Volumes, vol. 7, no. 1 (2013), pp. 664–669

[45] S. Digiesi, G. Mossa, S. Rubino, *Sustainable Order Quantity of Repairable Spare Parts*. Advanced Maintenance Engineering, IFAC Proceedings Volumes, vol. 2, no. 1 (2013), pp. 181–186

[46] S. Digiesi, G. Mossa, S. Rubino, A sustainable EOQ model for repairable spare parts under uncertain demand. IMA J. Manage. Math. **26**(2), 185–203 (2015)

[47] S. Digiesi, G. Mossa, G. Mummolo, Supply lead time uncertainty in a sustainable order quantity inventory model. Manage. Prod. Eng. Rev. **4**(4), 15–27 (2013)

[48] UmweltBundesamt, Probas database [Online]. Available: http://www.probas.umweltbundesamt.de/. Accessed 2009, 2012

[49] ISPRA, Sinanet database [Online]. Available: http://www.sinanet.isprambiente.it/. Accessed 2012

[50] B. Sahin, H. Yilmaz, Y. Ust, A. Guneri, B. Gulsun, An approach for analysing transport costs and a case study. Eur. J. Oper. Res. **193**(1), 1–11 (2009)

[51] M. Janic, Modelling the full costs of an intermodal and road freight transport network. Transport Res. D **12**(1), 33–44 (2007)

[52] R. Shi, Z. Hu, Y. Zhou, P. Liu, Research on railway freight market share based on the whole process of transport. Soc. Behav. Sci. **138**(1), 298–304 (2014)

[53] H. Lindstad, B.E. Asbjørnslett, E. Jullumstrø, Assessment of profit, cost and emissions by varying speed as a function of sea conditions and freight market. Transport Res. D **19**(1), 5–12 (2013)

[54] C. Schreyer, C. Schneider, M. Maibach, W. Rothengatter, C. Doll, D. Schmedding, *External Cost of Transport—Update Study* (International Railway Union, Paris, 2004)

[55] H. van Essen, A. Schroten, M. Otten, D. Sutter, C. Schreyer, R. Zandonella, M. Maibach, C. Doll, *External Costs of Transport in Europe—Update Study for 2008* (CE Delft, Delft, 2011)

[56] EU, *Roadmap to a Single European Transport Area—Towards A Competitive and Resource Efficient Transport System* (EC, Brussels, 2011)

[57] EU, *Strategy for Internalisation of External Costs, Communication from the Commission* (EC, Brussels, 2008)

[58] Society of Environmental Toxicology and Chemistry (SETAC), Guidelines for Life-Cycle Assessment: A 'Code of Practice' (Sesimbra, Portugal, 1993)

3 SOQ Model Formulations

Abstract
This Chapter, starting from the definition of the logistic cost function defined in Chap. 2, presents the Sustainable Order Quantity (SOQ) model formulation considering product demand and supply lead time firstly as deterministically known and then characterized by stochastic variability. In the stochastic problem, solution procedures are suggested for solving the SOQ problem. Five cases (numerical experiments and industrial full-scale) are presented to test the model. Sensitivity analyses are carried out varying costs, distances, and uncertainty of demand and lead time, thus showing the model capabilities and limitations. Finally, results are discussed and compared with classical EOQ solutions.

Keywords: Sustainable Order Quantity (SOQ), Stochastic variability of product demand, Stochastic variability of supply lead time, Sensitivity analysis, Automotive supply chain, Spare parts, Repair policy

3.1 Deterministic Demand and Lead Time

The following assumptions are assumed for the deterministic SOQ model formulation:

1. the annual demand is known;
2. the item consumption rate is constant over the year and deterministically known ($Q = G/H$);
3. transport is carried out by a single means of transport;
4. the transport cost per order and the transport lead time depend on the transport means adopted for a given route;
5. the supply lead time is calculated as the sum of transport time (T_T) and the time required for the material handling, order management and quality control (T_L); T_L is assumed to be a fraction ($k < 1$) of T_T; the transport speed is calculated from the free flow speed (v_{act}) of different transport means as $v = v_{act}/(1 + k)$; the value of the supply lead time is: $LT = L/v$.

In case of a deterministic demand and lead time, the total annual logistic cost function can be formulated as:

$$\Phi_L = c_h \cdot \frac{Q}{2} + \frac{G}{Q} \cdot c_0 + \frac{G}{Q} \cdot c_T(f) + \Phi_{EX} \quad [€/\text{year}]. \tag{3.1}$$

The optimal values of loss factor (f_{OPT}) and of the sustainable order quantity (SOQ) jointly minimizing the logistic and external costs can be obtained by solving problem (3.2):

S. Digiesi et al., *New Models for Sustainable Logistics*,
SpringerBriefs in Operations Management, DOI 10.1007/978-3-319-19710-4_3

$$\min_{SOQ,f} \Phi_L. \tag{3.2}$$

In case of deterministic demand and supply lead time, stock out events do not occur. For this reason, in (3.1) shortage costs are not considered. Under the assumptions made, order quantity (Q) can be obtained as:

$$Q = \frac{G \cdot L}{p \cdot H \cdot v}. \tag{3.3}$$

Parameter p is the ratio between the reorder level (r) and the order quantity (Q); it ranges in]0; 1] interval. For a given value of the average transport speed, the corresponding minimum value of Q can be evaluated assuming $p = 1$ ($r = Q$). This value represents the minimum lot size avoiding the stock out when a means of transport (and consequently a given average transport speed) has been selected (see Fig. 3.1).

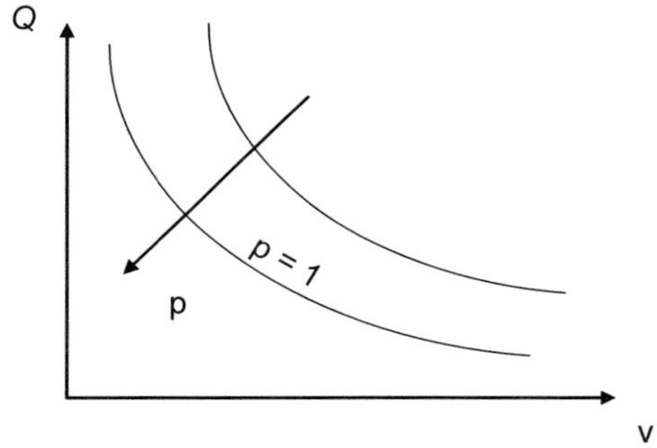

Fig. 3.1 Lot size (Q) vs. transport speed (v)

By substituting Eq. (3.3) into Eq. (3.1), the annual logistic cost function is:

$$\Phi_L = G \cdot \left[m \cdot c_T(f, L) + \frac{1}{2} \cdot \frac{c_h \cdot L}{p \cdot H \cdot v} \right] + \Phi_{EX} \quad [\text{€/year}]. \tag{3.4}$$

Starting from Eq. (3.4), a logistics, no-dimensional cost factor, F_L, is defined by dividing the total annual costs by the constant value $G \cdot c_h/2$; this value represents the annual inventory cost occurring in a limited situation where the order quantity is equal to the annual requirement (G):

$$F_L = \left\{ \frac{2 \cdot m}{c_h} \cdot c_T(f, L) + \frac{L}{p \cdot H \cdot v} \right\} + \frac{2 \cdot \Phi_{EX}}{G \cdot c_h}. \tag{3.5}$$

The general logistic optimization problem is finally defined as:

$$\min_{\{f,p\}} F_L. \tag{3.6}$$

The reorder level is, usually, assumed by the logistics management based on the lead time required from the order emission to the inventory replenishment. In such circumstances, f is the only optimization variable in the problem (3.6) since p values are assigned.

3.1.1 Environmental Costs

As discussed in Sect. 2, there are many categories of externalities due to the transport. In this Section, the emissions of transport and its related costs are considered in the logistics cost function. In case environmental costs of transport are considered in Eq. (3.1), the logistic cost function is obtained by substituting Eq. (2.32) and $E_R = E_S \cdot Q \cdot m \cdot L$ in Eq. (3.5).

$$F_L = \left\{ \frac{2 \cdot m}{c_h} \cdot c_T(f, L) + \frac{L}{p \cdot H \cdot v} \right\} + \frac{2 \cdot m \cdot L}{c_h} \sum_i \varepsilon_i \cdot e_i(f) \tag{3.7}$$

Starting from data in [1, 2] on specific energy consumption (E_s), the corresponding loss factors values (f) and emission factors (e_i) have been evaluated. Values obtained are in Table 3.1, where estimates of the transport speed adopted are also listed. Data adopted here are expressed in [$€_{2009}$].

Table 3.1 E_s, f, e_i, and v_{ACT} values for different means of transport

Means of transport			Es [TJ/t·km]	f [J/J]	$(CO_2)_{eq.}$ [kg/t·km]	$(SO_2)_{eq.}$ [kg/t·km]	$NMVOC_{eq.}$ [kg/t·km]	v_{act} [km/h]
Ship	oversee		9.44E-08	0.010	4.8E-02	4.7E-04	6.1E-04	70
Ship	internat.		1.00E-07	0.010	4.3E-02	4.8E-04	3.1E-04	46
Ship	coast ship		3.21E-07	0.033	9.4E-03	3.1E-04	2.7E-04	46
Rail	diesel		3.27E-07	0.033	8.7E-03	1.8E-04	1.3E-04	46
Rail	electric		3.60E-07	0.037	3.2E-02	2.8E-04	3.5E-04	70
Ship	inland		4.67E-07	0.048	8.3E-02	8.8E-04	4.1E-04	70
Truck	tir+append	32÷40 [t]	1.16E-06	0.118	6.1E-01	2.4E-03	3.1E-03	70
Truck	tir+append	28÷32 [t]	1.31E-06	0.134	4.1E-01	3.2E-03	5.1E-03	70
Truck	tir	20÷28 [t]	1.59E-06	0.162	2.2E-01	1.5E-03	2.6E-03	80
Truck	tir+append	< 28 [t]	1.67E-06	0.170	1.9E-01	1.4E-03	2.3E-03	80
Truck	tir	14÷20 [t]	2.10E-06	0.214	1.4E-01	1.1E-03	1.8E-03	80
Truck	tir	7.5÷14 [t]	2.37E-06	0.242	1.5E-01	9.7E-04	1.6E-03	80
Truck	tir	< 7.5 [t]	4.66E-06	0.475	1.2E-01	9.2E-04	1.5E-03	80
Truck	small tir	< 3.5 [t]	7.04E-06	0.718	1.0E-01	8.1E-04	1.3E-03	100
Airplane	internat.		9.70E-06	0.989	1.5E+00	5.5E-03	1.6E-03	700
Airplane	inland		1.74E-05	1.774	8.4E-01	3.0E-03	6.4E-04	700

A regression analysis carried out on data in Table 3.2 led to shape the dependency of the average speed of transport (v), obtained with a k value of 0.5, and of impact factors (e_i) on loss factor (f) by means of quadratic functions:

$$v = k_1 \cdot f^2 + k_2 \cdot f + k_3, \tag{3.8}$$

$$e_i = \gamma_i \cdot f^2 + \alpha_i \cdot f + \beta_i. \tag{3.9}$$

Regression parameters as well as R^2 values are in Table 3.2.

Table 3.2 Results of the regression analysis

		Regression parameters			
Impact category	Unit	α_i	β_i	γ_i	R^2
Global warming	$CO_{2eq.}$	0.8115	0.0097	0	0.9812
Acidification Potential	$SO_{2eq.}$	0.0027	0.0005	0	0.9182
Tropospheric Ozone Precursor Potential	$NMVOC_{eq.}$	0.0129	0.0000	−0.0123	0.9747
		k_1	k_2	k_3	
Transport speed	[km/h]	731.9	−370.6	65.4	0.8676

The unit transport costs have been modeled here by means of a quadratic function shaping the dependency of the costs on loss factor for different route lengths

$$c_T = a \cdot f^2 + b \cdot f + c \quad [\text{€/t}]. \tag{3.10}$$

The set of parameters (a,b,c) adopted for each distance value (L) are in Table 3.3.

Table 3.3 Parameters values of Eq. (3.10)

L	a	b	c
[km]	[€/t]	[€/t]	[€/t]
200	391.37	−402.35	108.69
300	370.86	−354.62	108.34
400	350.35	−306.9	107.99
500	329.84	−259.17	107.63
1000	227.28	−20.54	105.87
2000	22,168	456.72	102.35
3000	−182.94	933.98	98.82
4000	−388.06	1411.2	95.3
5000	−593.17	1888.5	91.77
10,000	−1618.7	4274.8	74.15

By adopting Eqs. (3.8), (3.9) and (3.10), the logistics cost function can be reformulated as:

$$F_L = \left\{ \begin{array}{l} \dfrac{2 \cdot m}{c_h}\left[a \cdot f^2 + b \cdot f + c\right] + \dfrac{1}{p \cdot H} \cdot \dfrac{L}{\left[k_1 \cdot f^2 + k_2 \cdot f + k_3\right]} \\ + \dfrac{2 \cdot m \cdot L}{c_h} \cdot \left[f^2 \cdot \sum_i \varepsilon_i \cdot \gamma_i + f \cdot \sum_i \varepsilon_i \cdot \alpha_i + \sum_i \varepsilon_i \cdot \beta_i \right] \end{array} \right\}. \tag{3.11}$$

As an example, the values of F_L and of its three terms (transport, inventory, and environmental costs, respectively) against the loss factor value, for given values of p (0.5), L (200 [km]) and c_h (5000 [€/t year]) are plotted in Fig. 3.2. As it can be observed, it is possible to identify a proper f value allowing minimizing both logistics and environmental cost.

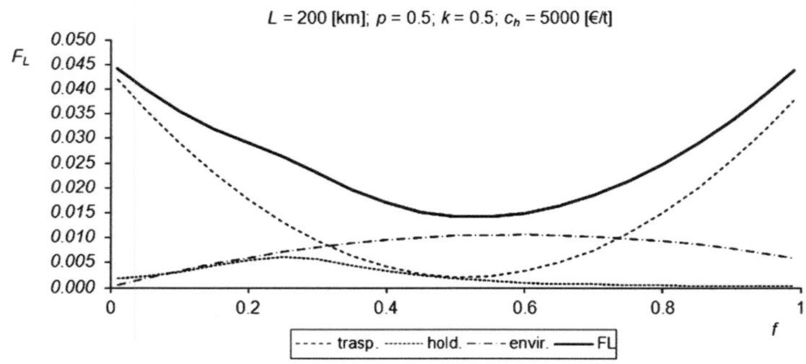

Fig. 3.2 Transport, environmental, holding costs and logistic cost factor F_L

In case of $p = 0.5$ and $c_h = 5000$ [€/t year], problem (3.6) has been solved for different L values (200–400–1000–2000 [km]). Results are in Fig. 3.3. As it can be observed, the optimal loss factor value (the loss factor which minimizes the global logistic function) decreases from about 0.6 to less than 0.1 with the increase of the route lengths from 200 [km] to over 1000 [km]. It can be concluded that in case of long route lengths, slow transport means, characterized by an average speed less than 50 [km/h], reveal a sustainable choice. In case of L > 500 [km], different transport means (rail: $f = 0.035$, or ships: $f = 0.010$) are characterized by approximately the same global logistic costs. On the contrary, in case of short distances ($L < 500$ [km]), a proper means of transport (tir < 7.5 [t]: $f = 0.475$, small tir < 3.5 [t]: $f = 0.718$) allows minimizing problem (3.11). Finally, it can be observed how transport by airplane ($f = 0.989$) does not represent a sustainable choice for any distance considered, since F_L assumes the highest value for any route length.

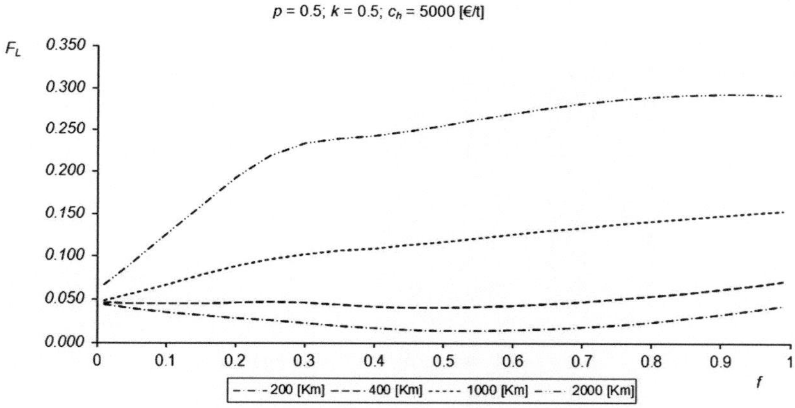

Fig. 3.3 F_L values for different route lengths and f values

As discussed above, the solution of problem (3.6) allows identifying the optimal loss factor value (f_{OPT}) and hence the optimal transport means for a given transport distance L. Each transport means is characterized by a proper average transport speed $v^*(f_{OPT})$.

By introducing $v^*(f_{OPT})$ in Eq. (3.3), the Sustainable Order Quantity, *SOQ*, minimizing the overall logistic cost function (Eq. 3.1), is obtained as:

$$SOQ = \frac{G \cdot L}{p \cdot H \cdot v^*(f_{OPT})}. \qquad (3.12)$$

Results of a sensitive analysis carried on *SOQ/G* ratio and f_{OPT} values varying the transport distance L, for different p values ($H = 3520$ [h/year]; $k = 0.5$; $c_h = 5000$ [€/t]) are in Fig. 3.4. $(SOQ/G)^{-1}$ ratio represents the number of annual orders.

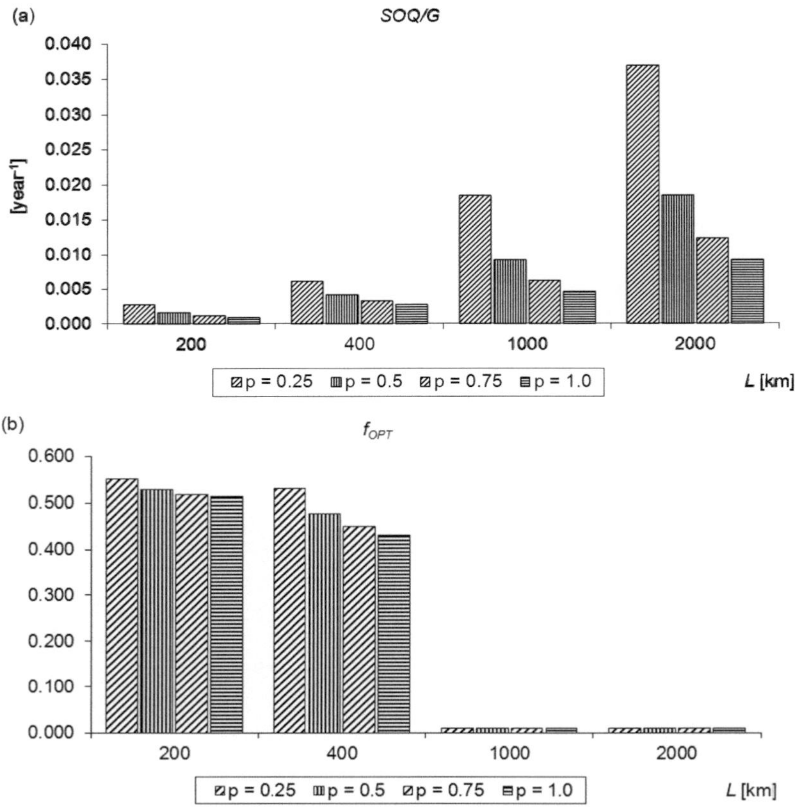

Fig. 3.4 *SOQ/G* (**a**) and f_{OPT} (**b**) versus transport distance L for different p values

It can be observed how, according to a sustainable logistics policy, in case of transport distance of more than 500 [km], slow transport means (low f_{OPT} values) and high order quantity should be adopted; in case of route lengths shortest than 500 [km], the solution move towards small SOQ/G values (see Fig. 3.4a) with medium average speed transport means (e.g. light trucks) (see Fig. 3.4b).

Parameter p (see Fig. 3.) represents the ratio between the supply lead time (LT) and the consumption time (CT). It proves to slightly affect f_{OPT} values (see Fig. 3.4a): even in the case of small supply lead time ($p = 0.25$), the proposed model does not identify high speed transport means as a sustainable choice because of the high environmental costs. On the contrary, SOQ/G value increases with the decreasing of the allowed (maximum) supply lead time/consuming time ratio.

Model Application

For given values of the route length L and of the holding cost c_h, problem (3.6) can be solved for different p values. The model provides the optimal solution of the logistics problem that is the couple (*SOQ*; transport means) minimizing the total logistics costs. As an example, in Fig. 3.5a solutions obtained for different p values in case of a transport distance $L = 400$ [km] and of a unit holding cost value $c_h = 5000$ [€/t·year] are showed. For each p value considered, the model identifies in trucks with different capacity (C) the optimal solution. In Fig. 3.5a, the corresponding SOQ values and the capacity of the truck (C) identified as optimal transport means are summarized. In case of $L = 440$ [km] the optimal solution consists of a truck with $C < 7.5$ [t], in the optimal order quantity $SOQ = 0.0041 \cdot G$, and in a p value of 0.5, that is in a reorder quantity $r = 0.5 \cdot SOQ$.

On the contrary, in case of $L = 1000$ [km], the optimal means of transport identified for any p value is a ship ($f = 0.010$); however, different SOQ values are obtained varying parameter p (see Fig. 3.5b).

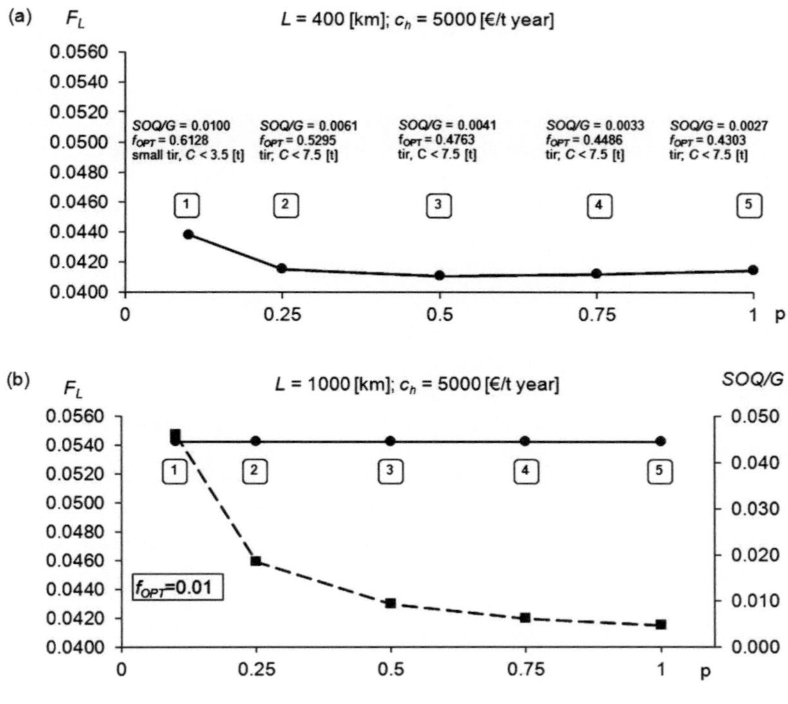

Fig. 3.5 F_L, SOQ/G, and f_{OPT} values for $c_h = 5000$ [€/t·year] in case of (**a**) short distances ($L = 400$ [km]) and (**b**) long distances ($L = 1000$ [km])

In order to assess the feasibility of the theoretical solution identified by the model, for each p value a final check is required: the allowed supply lead time [$LT_{ALL} = L/v(f_{OPT})$], has to be greater than the average supply lead time (LT), being LT evaluated on the basis of the average speed of the transport means selected ($LT = L / v$).

As an example, in case of $p = 0.5$, the limit (allowed) values of the supply lead time are listed in Table 3.4 for different route lengths ($k = 0.5$; $c_h = 5000$ [€/t·year]).

Table 3.4 Allowable lead time (LT_{ALL}) values obtained by the model in case of $p = 0.5$ ($k = 0.5$; $c_h = 5000$ [€/t·year])

	L [km]					
	200	300	400	500	1000	2000
LT_{ALL} [h]	2.5	4.4	6.6	8.1	16.2	32.4

In case of no transport means available satisfies the above constraint, three strategies could be adopted.

(a) Adoption of a different reorder level, that is of a different sub-optimal solution; generally, this entails an increase in the logistic costs as per Fig. 3.5a; in case of long transport distance (L > 500 [km], see Fig. 3.5b), however, negligible differences are observed; as an example, in case of $L = 400$ [km], the sub-optimal solution 4 satisfies the lead time constraint ($LT_{ALL} < LT$) with an increase of less than 1 % in the logistic costs, as per data in Table 3.4.

(b) In case no sub-optimal solutions are available, a faster available means of transport has to be selected with the result of an increase in the logistic costs; as an example, in case of $L = 300$ [km], it is not possible to find a feasible solution (see Table 3.5). For $p = 0.5$, by adopting a means of transport that is faster than the one identified by the model (solution *3mod.* in Table 3.6) it is possible to satisfy the time constraint ($LT_{ALL} < LT$), but this leads to an increase in the total logistic costs of about 30 %.

(c) In the case that there is no faster means of transport available, in order to satisfy the time constraint a lower k value is required, that is material handling, order management, and the quality control time have to be reduced. Generally, low decreases in k are possible.

Table 3.5 Solutions of the logistics problem (36) in case of L = 400 [km] (k = 0.5; c_h = 5000)

Solution	p	SOQ/G	LT_{ALL} [h]	f_{OPT}	Transport means	LT [h]	F_L
1	0.10	0.0100	3.54	0.613	Truck; C < 3.5[t]	6.00	0.0438
2	0.25	0.0061	5.38	0.530	Truck; C < 7.5[t]	7.50	0.0415
3	0.50	0.0041	7.29	0.476	Truck; C < 7.5[t]	7.50	0.0411
4	0.75	0.0033	8.62	0.449	Truck; C < 7.5[t]	7.50	0.0412
5	1.00	0.0027	9.66	0.430	Truck; C < 7.5[t]	7.50	0.0415

Table 3.6 Solutions of the logistics problem (36) in case of L = 300 [km] (k = 0.5; c_h = 5000 [€/t·year])

solution	p	SOQ/G	LT_{ALL} [h]	f_{OPT}	transport means	LT [h]	F_L
1	0.10	0.0078	2.74	0.606	truck; C < 3.5[t]	4.50	0.0293
2	0.25	0.0043	3.81	0.540	truck; C < 7.5[t]	5.63	0.0280
3	0.50	0.0025	4.48	0.502	truck; C < 7.5[t]	4.48	0.0277
3mod.	*0.50*	*0.0025*	*4.48*	*0.7176**	*truck; C < 3.5[t]*	*4.48*	*0.0357*
4	0.75	0.0020	5.21	0.485	truck; C < 7.5[t]	5.63	0.0278
5	1.00	0.0016	5.54	0.474	truck; C < 7.5[t]	5.63	0.0279

3.1.2 Environmental and Social Costs

The main goal of the model application proposed in this Section is to evaluate how different internalization strategies can promote the shift from the adoption of transport means generating high external costs toward transport means characterized by better environmental and social performances.

Starting from Eq. (3.1) and Eq. (3.3) and explicitly considering the order costs, the annual logistics cost function can be evaluated as:

$$\Phi_L = G \cdot \left[m \cdot c_T(f, L) + \frac{1}{2} \cdot \frac{c_h \cdot L}{p \cdot H \cdot v} + c_O \cdot \frac{p \cdot H \cdot v}{L} \right] + \Phi_{EX}. \tag{3.13}$$

Table 3.7 summarizes the unit external costs adopted in this Section: cost data for water, rail and road freight transport come from [3], cost data for air freight transport comes from [4]. Unit external costs of road transport are provided for both LDV (Light Duty Vehicles) category, with a GWR (Gross Weight Rate) less than 3.5 [t], and HDV (Heavy Duty Vehicles) category, with a GWR greater than 3.5 [t]. The LCA category considers the up- and down processes external costs. Cost data have been updated to [$€_{2013}$/t·km] values by means of discounting indexes available in [5].

Table 3.7 Unit external costs [$€_{2013}$/t·km] of different transport means

Cost category	Water-borne	Rail electric	Rail diesel	LDV	HDV	Airplane
			[$€_{2013}$/t·km]			
Accidents	0.0	0.2	0.2	62.3	11.3	0.0
Air Pollution	6.0	1.0	1.9	19.8	7.4	20.7
Noise	0.0	1.1	1.1	7.0	2.0	11.8
Congestion	0.0	0.0	0.0	107.0	28.5	0.0
GW	4.0	0.0	4.3	49.3	10.9	325.9
LCA	1.4	4.4	5.7	15.8	3.3	10.2
Other external cost	1.0	0.6	0.6	7.1	2.8	5.0
Overall	*12.4*	*7.3*	*13.7*	*268.3*	*66.2*	*373.7*

By adopting average values of the loss factor for the different transport modalities in Table 3.7, it is possible to shape the dependency of the overall unit external costs on the loss factor by means of a quadratic function:

$$\varepsilon_{EX} = \delta \cdot f^2 + \mu \cdot f. \tag{3.14}$$

Being:

$$\Phi_{EX} = \varepsilon_{EX}(f) \cdot m \cdot G, \tag{3.15}$$

by adopting quadratic functions to shape the dependency of the speed (see Eq. 3.8) and of the unit transport costs (see Eq. 3.10) on the loss factor as in the previous Section, the logistics cost factor F_L is:

$$F_L = \left\{ \begin{array}{l} \frac{2}{c_h} \cdot m \cdot \left[a \cdot f^2 + b \cdot f + c \right] + \frac{1}{p \cdot H} \cdot \frac{L}{\left[k_1 \cdot f^2 + k_2 \cdot f + k_3 \right]} + \\ \frac{2 \cdot c_O \cdot p \cdot H}{G \cdot c_H \cdot L} \cdot \left[k_1 \cdot f^2 + k_2 \cdot f + k_3 \right] + \frac{2 \cdot m \cdot L}{c_h} \cdot \left[\delta \cdot f^2 + \mu \cdot f \right] \end{array} \right\}. \tag{3.16}$$

A Numerical Example

In this case study, updated data on loss factor (f) values as well as on transport speed (v) have been adopted. Data are in Table 3.8.

Table 3.8 Loss factor (f) and average speed of transport (v) values for different means of transport; [DB1] = [1], 2012; [DB2] = [6], 2012; [DB3] = [7], 2007; [DB4] = [8], 2012

Transport means	Database	f	Database	v [km/h]
Ship—oversee	[DB1]	0.010	[DB3]	24.7
Ship—tank	[DB1]	0.021	[DB3]	18.5
Ship—coast	[DB1]	0.033	[DB3]	17.3
Rail—diesel	[DB1]	0.033	Assumed	46.7
Rail—electric	[DB1]	0.037	Assumed	46.7
Ship—inland	[DB1]	0.048	[DB3]	24.7
Truck 34 ÷ 40 [t]	[DB1], [DB2]	0.063	[DB4]	56.5
Truck 28 ÷ 34 [t]	[DB1], [DB2]	0.069	[DB4]	56.5
Truck 26 ÷ 28 [t]	[DB1], [DB2]	0.119	[DB4]	56.5
Truck 20 ÷ 26 [t]	[DB1], [DB2]	0.136	[DB4]	56.5
Truck 14 ÷ 20 [t]	[DB1], [DB2]	0.127	[DB4]	64.0
Truck 12 ÷ 14 [t]	[DB1], [DB2]	0.139	[DB4]	64.0
Truck 7.5 ÷ 12 [t]	[DB1], [DB2]	0.149	[DB4]	64.0
Truck 3.5 ÷ 7.5 [t]	[DB1], [DB2]	0.165	[DB4]	64.0
Truck 3.5 ÷ 7.5 [t][a]	[DB1], [DB2]	0.156	[DB4]	64.0
Truck <3.5 [t][b]	[DB1], [DB2]	0.335	[DB4]	72.7
Truck <3.5 [t][c]	[DB1]	0.554	[DB4]	72.7
Airplane	[DB1]	0.984	Assumed	466.7

[a]Truck operated without append
[b]Light good vehicles
[c]Vans

Data in [8] on free flow vehicle speeds observed in UK in 2011 have been elaborated in order to obtain an average speed value for each truck category (see Table 3.9). The actual speed value (v_{ACT}) for each truck category has been obtained from observed data by weighing the values observed on the different road types (motorways, dual carriageways, single carriageways) with the corresponding observations number. The related average value of the transport speed has been obtained using a k factor of 0.5.

Data on ships cruise speeds as in [7] have been adopted. For railroads and aircraft transport means, a cruise speed of 70 [km/h] and 700 [km/h] have been assumed, respectively.

Table 3.9 UK statistics adopted (year 2011)

		LDV	HDV				
		<3.5 [t]	3.5–18 [t]	26 [t]	30–44 [t]	26–40 [t]	40–44 [t]
Motorways	Average speed [mph]	69.54	61.09	54.20	53.57	53.84	53.61
	Observations [# × 10^3]	82884	6369	2863	1716	7260	42830
Dual carriageways	Average speed [mph]	67.99	59.48	53.46	52.99	53.01	53.09
	Observations [# × 10^3]	7127	2548	257	210	453	2762
Single carriageways	Average speed [mph]	48.09	46.20	42.41	42.76	42.72	44.24
	Observations [# × 10^3]	5816	1957	223	167	275	1164
	v_{ACT} [mph]	68.12	58.03	53.36	52.65	53.41	53.35
	v_{ACT} [km/h]	108.99	92.85	85.37	84.24	85.45	85.36

By relating the loss factor values of different transport means with their characteristic average speed in transport (obtained assuming $k = 0.5$), the data set in Table 3.8 has been obtained. Results of a regression analysis carried out on the data set showed that the relationship between the speed of transport and the loss factor value of the transport means adopted is best shaped by a quadratic function:

$$v = k_1 \cdot f^2 + k_2 \cdot f + k_3, \tag{3.17}$$

with $k_1 = 557.6$; $k_2 = -150.4$; $k_3 = 52.7$ ($R^2 = 0.935$).
Starting from the data set in Table 3.8, average values of the loss factor for the different transport modalities have been evaluated. Values adopted are in Table 3.10.

Table 3.10 Average values of the loss factor for different transport modalities

Transport means	f
Sea Ship	0.021
Diesel Train	0.033
Electric Train	0.037
Inland Ship	0.048
HDV	0.125
LDV	0.444
Airplane	0.984

A regression analysis carried out on data in Table 3.7 led to evaluate parameters values of Eq. (3.14) as: $\delta = -607.17$, $\mu = 979.22$ ($R^2 = 0.9718$).
Table 3.11 summarizes the data set adopted for the numerical experiment.

In order to evaluate the optimal transport means and the optimal lot-size value minimizing the logistics function F_L, problem (3.6), considering also Eq. (3.3), has been solved. For each transport distance considered both the solution obtained in case of internalization of the whole external costs ($F_{L,SUST}$ in Table 3.12, SOQ_{SUST} in Table 3.13 and r_{SUST} in Table 3.14) and the solution obtained in case of no internalization of external costs ($F_{L,ECON}$ in Table 3.12, SOQ_{ECON} in the Table 3.13and r_{ECON} in Table 3.14) are showed. In brackets, the optimal transport means is specified (IV = Inland Vessel, ET = Electric Train, SS = Sea Ship).

The optimal lot size and the reorder level, according to the model defined, proved to be independent on the external costs. For each transport means, *SOQ* and r values depend only on the transport distance considered (see Table 3.15 and Table 3.16).

Table 3.11: Data set adopted for the numerical experiment

Parameters	Values	Measurement units
G	9224	[unit/year]
H	3520	[h/year]
m	0.5	[kg]
L	200; 300; 400; 500; 1000; 2000; 3000; 4000; 5000; 10,000	[km]
c_H	4.65	[€/unit·year]
	See Eq. (3.10) and Table	
c_T	3.3	[€/t]
c_O	1	[€/order]

Sensitivity Analysis
In order to evaluate the impact of different internalization strategies on specific logistics costs [€/kg·km], the problem (3.6) has been solved for different extra costs charged in the logistics costs function. Ten different internalization strategies have been considered, with a percentage of charged extra costs ranging between 0 % and 200 % of the whole external costs (100 %-case values are in Table 3.7). The specific logistics cost is defined here as:

$$\phi_L = \frac{\Phi_L}{G \cdot m \cdot L} \quad . \tag{3.18}$$

Starting from solutions obtained by the model, the increase of the specific logistics cost (compared with the base case of zero external costs charged) have been computed for each case. Values are in Table 3.17, where in brackets there is the optimal transport means identified by the model in each case.

Table 3.12 $F_{L,ECON}$ and $F_{L,SUST}$ for different transport distances

	200	300	400	500	1000	2000	3000	4000	5000	10,000
	[km]									
$F_{L,ECON}$	0.015 (LDV)	0.019 (LDV)	0.022 (LDV)	0.026 (LDV)	0.036 (IV)	0.04 (SS)	0.045 (SS)	0.052 (SS)	0.058 (SS)	0.093 (SS)
$F_{L,SUST}$	0.027 (LDV)	0.033 (HDV)	0.035 (IV)	0.036 (IV)	0.039 (ET)	0.047 (ET)	0.053 (SS)	0.062 (SS)	0.072 (SS)	0.12 (SS)

Table 3.13 SOQ_{ECON} and SOQ_{SUST} for different transport distances

	200	300	400	500	1000	2000	3000	4000	5000	10,000
	[km]									
SOQ_{ECON}	65 (LDV)	61 (LDV)	62 (LDV)	64 (LDV)	62 (IV)	105 (SS)	158 (SS)	210 (SS)	263 (SS)	526 (SS)
SOQ_{SUST}	65 (LDV)	63 (HDV)	62 (IV)	62 (IV)	63 (ET)	109 (ET)	158 (SS)	210 (SS)	263 (SS)	526 (SS)

Table 3.14 r_{ECON} and r_{SUST} for different transport distances

	200	300	400	500	1000	2000	3000	4000	5000	10,000
	[km]									
r_{ECON}	5 (LDV)	8 (LDV)	11 (LDV)	14 (LDV)	56 (IV)	105 (SS)	158 (SS)	210 (SS)	263 (SS)	526 (SS)
r_{SUST}	5 (LDV)	18 (HDV)	22 (IV)	28 (IV)	55 (ET)	109 (ET)	158 (SS)	210 (SS)	263 (SS)	526 (SS)

Table 3.15 Sustainable Order Quantity (SOQ) for different transport means and distances

Transport means	200	300	400	500	1000	2000	3000	4000	5000	10,000
	[km]									
Sea Ship	63	64	63	62	64	105	158	210	263	526
Diesel Train	62	62	62	62	62	109	163	217	271	542
Electric Train	63	62	63	63	63	109	164	219	273	547
Inland Vessel	62	62	62	62	62	112	168	224	280	560
HDV	62	63	63	64	64	123	185	246	307	615
LDV	65	61	62	64	64	64	82	109	136	273
Airplane	62	63	64	64	64	64	63	62	62	64

Table 3.16 Reorder level (r) for different transport means and distances

Transport means	200	300	400	500	1000	2000	3000	4000	5000	10,000
	[km]									
Sea Ship	11	16	21	26	53	105	158	210	263	526
Diesel Train	11	16	22	27	54	108	163	217	271	542
Electric Train	11	16	22	27	55	109	164	219	273	547
Inland Vessel	11	17	22	28	56	112	168	224	280	560
HDV	12	18	25	31	61	123	184	246	307	615
LDV	5	8	11	14	27	55	82	109	136	273
Airplane	1	2	2	3	6	12	18	24	29	59

Table 3.17 Percentage increase of the specific logistics cost for different internalization strategies compared to the economic case (External costs Charging level = 0 %)

Carging level	[km] 200	300	400	500	1000	2000	3000	4000	5000	10,000
0 %	0 %(LDV)	0 %(LDV)	0 %(LDV)	0 %(LDV)	0 %(IV)	0 %(SS)	0 %(SS)	0 %(SS)	0 %(SS)	0 %(SS)
20 %	15 %(LDV)	18 %(LDV)	21 %(LDV)	22 %(LDV)	1 %(ET)	3 %(SS)	4 %(SS)	4 %(SS)	5 %(SS)	6 %(SS)
40 %	30 %(LDV)	37 %(LDV)	41 %(LDV)	30 %(HDV)	2 %(ET)	5 %(SS)	7 %(SS)	8 %(SS)	9 %(SS)	11 %(SS)
60 %	46 %(LDV)	55 %(LDV)	48 %(HDV)	35 % (IV)	3 %(ET)	8 %(SS)	11 %(SS)	12 %(SS)	14 %(SS)	17 %(SS)
80 %	61 %(LDV)	71 %(HDV)	53 %(HDV)	36 %(IV)	3 %(ET)	11 %(SS)	14 %(SS)	17 %(SS)	18 %(SS)	23 %(SS)
100 %	76 %(LDV)	75 %(HDV)	56 %(IV)	37 % (IV)	4 %(ET)	13 %(ET)	18 %(SS)	21 %(SS)	23 %(SS)	29 %(SS)
120 %	91 %(LDV)	80 %(HDV)	57 %(IV)	37 %(ET)	5 %(ET)	14 %(ET)	21 %(ET)	25 %(SS)	27 %(SS)	34 %(SS)
140 %	106 %(LDV)	84 %(IWT)	58 %(IV)	38 %(ET)	6 %(ET)	16 %(ET)	23 %(ET)	28 %(ET)	31 %(ET)	40 %(SS)
160 %	111 %(HDV)	85 %(IWT)	59 %(ET)	39 %(ET)	7 %(ET)	17 %(ET)	25 %(ET)	30 %(ET)	34 %(ET)	45 %(ET)
180 %	115 %(HDV)	86 %(IWT)	59 %(ET)	39 %(ET)	8 %(ET)	19 %(ET)	27 %(ET)	33 %(ET)	37 %(ET)	48 %(ET)
200 %	119 %(HDV)	87 %(IWT)	60 %(ET)	40 %(ET)	9 %(ET)	21 %(ET)	29 %(ET)	35 %(ET)	39 %(ET)	51 %(ET)

For the shortest transport distance considered (200 [km]) road transport is the optimal solution for each of the internalization strategies considered. An external costs charging, in the logistics function, of more than 160 % of the whole external cost generated from the transport causes a shift from LDV to HDV transport modality.

In case of short distances ($L = 300$, 400, and 500 [km]), road transport means are the optimal solution of the problem not for all the internalization strategies considered. Depending on the transport distance, the increase of the percentage of the whole external costs charged in the logistics cost function promotes the shift toward more sustainable transport modes (inland vessel, electric train).

Road transport is not solution of the problem (3.6) in case of transport distances longer than 1000 [km] for all the internalization strategies considered.

Fig. 3.6 shows the trend of the specific logistics costs for all the transport means considered in case of four different distances ($L = 200$, 500, 1000, 2000 [km]) and three different internalization strategies (0 %, 100 %, and 200 %). In Fig. 3.7 the specific logistics costs for different transport distance (L) and different internalization policies adopted are depicted.

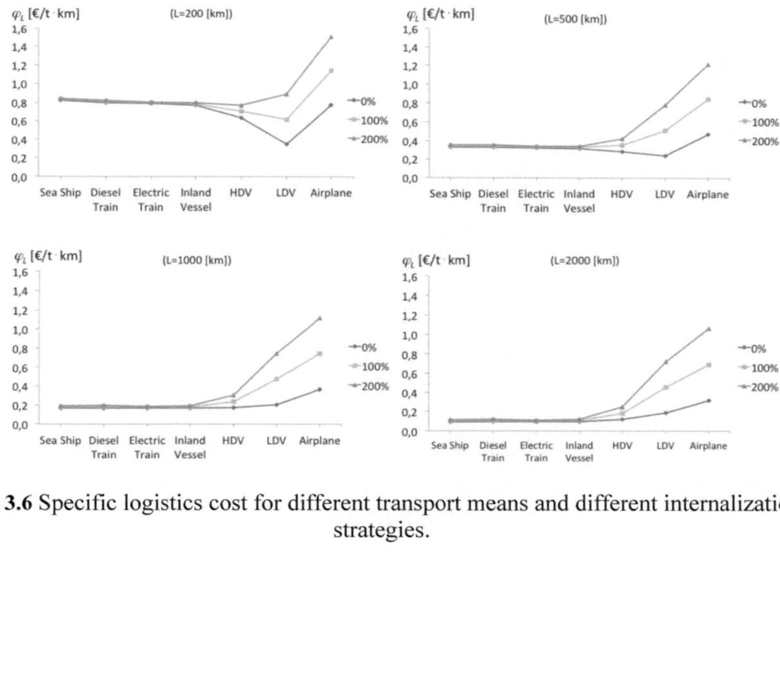

Fig. 3.6 Specific logistics cost for different transport means and different internalization strategies.

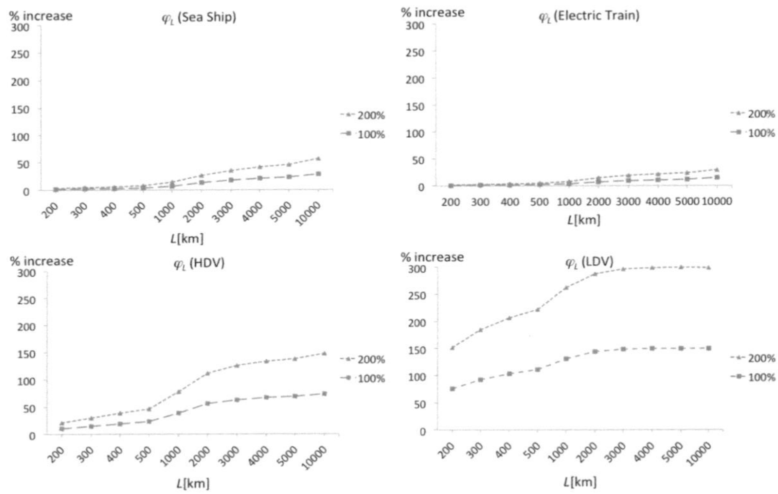

Fig. 3.7 Specific logistics cost percentage increase compared to the economic case for different transport distance (L) and two different internalization strategies including all the external costs categories

As shown in Fig. 3.7, for a given distance, the adoption of a more severe internalization strategy results in an increase of the specific logistics cost which significantly differs depending on the transport means considered. As an example, for a transport distance of 1000 [km] and with an increase from 100 % to 200 % of the percentage of external costs charged, the increase in the specific logistics cost is below of the 8 % for the sea ship and electric train transport mode. On the contrary, substantial differences have been observed in case of road transport modalities: 28 % in case of HDV and 57 % in case of LDV.

GW and LCA costs assume the larger values in the economic evaluation of the external costs of transport: the sum of these two categories weights up to 90 % (air transport) of the total external costs. Moreover, in the evaluation of sustainability in the transport sector, often only the carbon emissions are considered. A sensitivity analysis has been carried out in order to evaluate the effects of different internalization strategies if only GW and LCA external costs are charged in the logistics cost function. Results obtained are in Table 3.18 and in Fig. 3.8.

Table 3.18 Percentage increase of the specific logistics cost in case of different internalization strategies charging only GW and LCA external costs categories compared to the economic case (External costs charging level = 0 %)

Carging level	[km] 200	300	400	500	1000	2000	3000	4000	5000	10,000
0 %	0 %(LDV)	0 %(LDV)	0 %(LDV)	0 % (LDV)	0 % (IV)	0 %(SS)	0 %(SS)	0 %(SS)	0 %(SS)	0 % (SS)
20 %	4 %(LDV)	4 %(LDV)	5 %(LDV)	5 % (LDV)	1 % (ET)	1 %(SS)	2 %(SS)	2 %(SS)	2 %(SS)	3 % (SS)
40 %	7 %(LDV)	9 %(LDV)	10 %(LDV)	11 % (LDV)	1 % (ET)	2 %(SS)	3 %(SS)	4 %(SS)	4 %(SS)	5 % (SS)
60 %	11 %(LDV)	13 %(LDV)	15 %(LDV)	16 % (LDV)	2 % (ET)	4 %(SS)	5 %(SS)	5 %(SS)	6 %(SS)	8 % (SS)
80 %	15 %(LDV)	18 %(LDV)	20 %(LDV)	22 %(LDV)	2 %(ET)	5 %(SS)	6 %(SS)	7 %(SS)	8 %(SS)	10 %(SS)
100 %	18 %(LDV)	22 %(LDV)	25 %(LDV)	25 %(HDV)	3 %(ET)	6 %(SS)	8 %(SS)	9 %(SS)	10 %(SS)	13 %(SS)
120 %	22 %(LDV)	27 %(LDV)	30 %(LDV)	26 %(HDV)	3 %(ET)	7 %(SS)	9 %(SS)	11 %(SS)	12 %(SS)	15 %(SS)
140 %	26 %(LDV)	31 %(LDV)	35 %(LDV)	27 %(HDV)	4 %(ET)	8 %(SS)	11 %(SS)	13 %(SS)	14 %(SS)	18 %(SS)
160 %	29 %(LDV)	36 %(LDV)	40 %(LDV)	28 %(HDV)	4 %(ET)	9 %(SS)	12 %(SS)	14 %(SS)	16 %(SS)	20 %(SS)
180 %	33 %(LDV)	40 %(LDV)	43 %(HDV)	29 %(HDV)	5 %(ET)	11 %(SS)	14 %(SS)	16 %(SS)	18 %(SS)	23 %(SS)
200 %	37 %(LDV)	45 %(LDV)	44 %(HDV)	31 %(HDV)	5 %(ET)	12 %(SS)	15 %(SS)	18 %(SS)	20 %(SS)	25 %(SS)

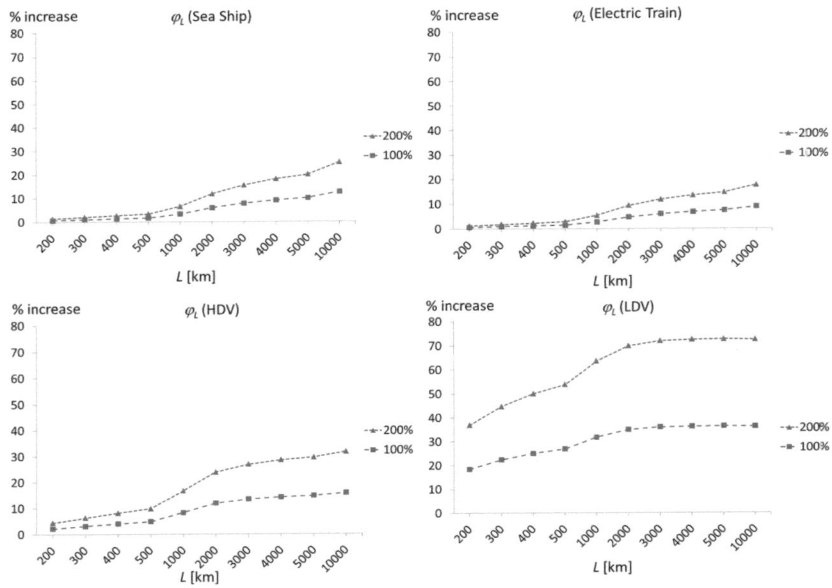

Fig. 3.8 Specific logistics cost percentage increase compared to the economic case for different transport distance (L) and two different internalization startegies charging only GW and LCA external costs categories.

The trends in Fig. 3.8 are similar to those of Fig. 3.7 but with a smaller specific logistics costs increase (8 % in case of HDV, 24 % in case of LDV and below the 4 % in case of electric train and the sea ship).

The internalization strategies charging only the costs of GW and LCA of transport reveal to be ineffective in terms of more environmental and social friendly transport means selection in case of short transport distances ($L \leq 500$ [km], see Table 3.18).

3.2 Stochastic SOQ Model

The formulation of the SOQ model proposed in Sect. 3.1 allows identifying the optimal lot-size and the transport means minimizing logistic and environmental costs of transport in case of a deterministic product demand. In case of stochastic variability of product demand and/or supply lead time, shortage costs, as well as safety stock costs, have to be considered in the logistics cost function (see Eq. 3.19).

$$\Phi_L = \Phi_O + \Phi_H + \Phi_T + \Phi_S + \Phi_{EX} \quad [\text{€/year}] \tag{3.19}$$

Usually, safety stock is sized searching for the trade-off between holding and shortage costs due to stock out events. In this Section, two extension of the SOQ model in case of stochastic variability of product demand and of supply lead time are discussed. Safety stock and shortage costs are considered in the logistics cost functions proposed. Analytical models provide order quantity, transport means and safety stock size allowing minimizing logistics and environmental costs in case of stochastic variability of product demand or of supply lead time. Results of the application of the models to a full scale case study from automotive industry are presented and discussed.

3.2.1 Product Demand Uncertainty

The following assumptions are assumed in defining the SOQ model in case of stochastic variability of the product demand:

1. the expected annual demand (G) is known;
2. the product demands in each time period, D_i, are independent stochastic variables;
3. in each i-th period of the lead time (LT), product demand is characterized by the same expected value and by the same standard deviation:

$$E(D_1) = E(D_2) = \ldots = E(D_{LT}) = E(D) = G/H, \tag{3.20}$$

$$\sigma_{D_1}^2 = \sigma_{D_2}^2 = \ldots = \sigma_{DLT}^2 = \sigma_D^2. \tag{3.21}$$

As far as concern the transport speed (v) and the evaluation of the lead time (LT), the same assumptions as in Sect. 3.1.1 are adopted.

In case of a product demand varying stochastically, and environmental costs of transport considered, the logistic cost function (3.19) can be written as:

$$\Phi_L = c_h \cdot \left[\frac{1}{2}Q + SS\right] + G \cdot m \cdot c_T + \frac{G}{Q} \cdot c_S \cdot N_S + G \cdot m \cdot L \cdot \sum_i \varepsilon_i \cdot e(f). \tag{3.22}$$

In Eq. (3.22) the safety stock level, SS, consistent with an assigned service level, SL, is evaluated as:

$$SS = LT \cdot \left[D^* - E(D)\right], \tag{3.23}$$

where:

$$SL = prob\left\{D_{TOT} \le LT \cdot D^*\right\} = \int_{-\infty}^{LT \cdot D^*} pdf(D_{TOT})dD_{TOT}. \tag{3.24}$$

The number of stock out events is computed (see Sect. 2.4.4) by means of

$$N_S = \int_{D^*_{TOT}}^{+\infty} pdf(D_{TOT}) \cdot (D_{TOT} - D^*_{TOT}) \cdot dD_{TOT}, \tag{3.25}$$

where

$$pdf(D_{TOT}) = \frac{1}{\sqrt{LT} \cdot \sigma_D \cdot \sqrt{2\pi}} \cdot \exp\left[-\frac{(D_{TOT} - LT \cdot E(D))^2}{2 \cdot LT \cdot \sigma_D^2} \right], \tag{3.26}$$

and

$$D_{TOT} = D_1 + D_2 + \ldots + D_{LT}. \tag{3.27}$$

In Eqs. (3.23)–(3.25) D^* and D^*_{TOT} are the maximum demand value of the i-th period and the maximum demand value on LT periods not causing a stock-out in that period and in LT periods, respectively. From Eq. (3.27) it is obtained: $LT \cdot D^* = D^*_{TOT}$.

The optimal values of loss factor (f_{OPT}), order quantity (SOQ), and the corresponding values of the reorder level ($r(f_{OPT})$), and of the safety stock ($SS(f_{OPT})$) jointly minimizing the logistics and external costs can be obtained by solving problem (3.28) or the equivalent problem (3.29):

$$\min_{SOQ,f} \Phi_L \tag{3.28}$$

$$\min_{p,f} F_L \tag{3.29}$$

By solving problem (3.27) or (3.28) the evaluation of the optimal loss factor value led to jointly identify the optimal order quantity (SOQ), and the optimal safety stock level (SS) as:

$$SOQ = \frac{G \cdot L}{p_{OPT} \cdot H \cdot v(f_{OPT})} \tag{3.30}$$

$$SS = \frac{L}{v(f_{OPT})} \cdot \left[D^*(f_{OPT}) - E(D) \right] \tag{3.31}$$

Case Study

The model has been applied to a case study from automotive industry [9, 10]. The case refers to a multi-site manufacturing system producing breaking equipment. The supply chain consists of three production sites (see Fig. 3.9): sites

1 and 2 are responsible of producing semi-finished products for sites 2 (P1C, P2C) and 3 (P1B, P2B); in site 3, three finite products (P1, P2, P3) are assembled for a single customer. Sites 2 and 3 produce additional products (PX, PY) required by external customers of the aftermarket starting from semi-finished product (PXf, PYf) externally supplied. The products demand of the main customer, as well as of the external customers, are uncertain and vary stochastically.

The logistics problem (3.28) has been solved for PXf product in site 1. The PXf product demand is characterized by an expected value (*E(D)*) of 2.62 [unit/h] and a coefficient of variation *cv* of 0.26. The number of working hours (*H*) in site 1 is 3520 [h/year]. The maximum production capacity of product PX in site 1 is 18.40 [unit/h]. This value is a theoretical limit for the production capacity, since it is obtained without considering the set-up time.

The unit holding cost of product PXf in site 1 (c_h) is 15.49 [€/unit], and the extra cost generated in case of stock out (c_S) is 10.06 [€/unit] ($c_S/c_h = 0.65$). The mass (m) of the product PXf is 0.5 [kg] (Fig. 3.9). In evaluating (3.23) and (3.24), a lower bound value of 0 [unit/h] and an upper bound value of 18.40 [unit/h] have been considered, respectively. The analysis has been limited to road transport because of geographical constraints. Due to unavailability of data, order costs have not been considered.

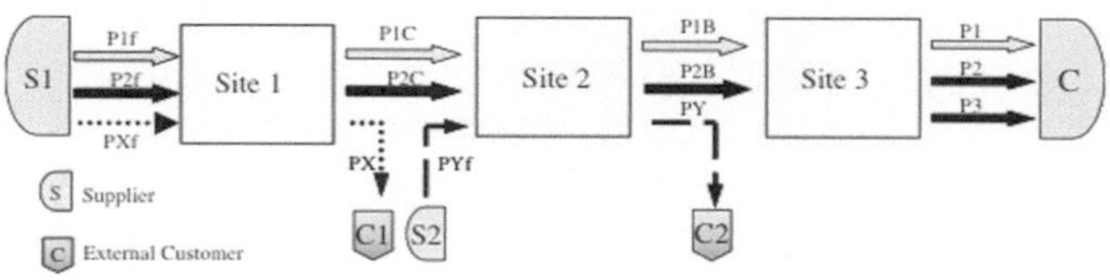

Fig. 3.9 The supply chain of a multi-site manufacturing system

In this case study, updated data on emission factors (e_i) have been adopted. Data are in Table 3.19.

Loss factor (*f*) and average transport speed (*v*) values adopted are those of Sect. 3.1.2. As in Sect. 3.1.2, average speed of transport has been shaped as a quadratic function of the loss factor:

$$v = k_1 \cdot f^2 + k_2 \cdot f + k_3 \tag{3.32}$$

with $k_1 = 557.6$; $k_2 = -150.4$; $k_3 = 52.7$.

Equivalent emission values in Table 3.19 have been obtained starting from data on emission factors of different transport means in [1, 6]. Data allowed evaluating the impacts on the environment in terms of Global Warming (GW), Acidification (AP) and Tropospheric Ozone Precursor Potential (TOPP) of different means of transport. Starting from the unit emission [kg/t·km] of different pollutants, the impacts have been measured by evaluating the equivalent emissions of CO_2 (GW),

SO_2 (AP), and NMVOC (TOPP). In case of GW and AP, classification factors as in [11] have been adopted; in case of TOPP classification factors as in [12] have been adopted (see Table 3.20).

Table 3.19 e_i values for different means of transport; [DB1] = [1], 2012; [DB2] = [6], 2012

Transport means	f	Database	$(CO_2)_{eq.}$ [kg/t·km]	$(SO_2)_{eq.}$ [kg/t·km]	$NMVOC_{eq.}$ [kg/t·km]
Ship—oversee	0.010	[DB1]	0.027	0.000	0.000
Ship—tank	0.021	[DB1]	0.018	0.001	0.000
Ship—coast	0.033	[DB1]	0.044	0.000	0.000
Rail—diesel	0.033	[DB1]	0.032	0.000	0.000
Rail—electric	0.037	[DB1]	0.083	0.001	0.000
Ship—inland	0.048	[DB1]	0.044	0.000	0.001
Truck 34 ÷ 40 [t]	0.063	[DB1], [DB2]	0.054	0.000	0.000
Truck 28 ÷ 34 [t]	0.069	[DB1], [DB2]	0.059	0.000	0.001
Truck 26 ÷ 28 [t]	0.119	[DB1], [DB2]	0.101	0.000	0.001
Truck 20 ÷ 26 [t]	0.136	[DB1], [DB2]	0.116	0.001	0.000
Truck 14 ÷ 20 [t]	0.127	[DB1], [DB2]	0.109	0.001	0.001
Truck 12 ÷ 14 [t]	0.139	[DB1], [DB2]	0.119	0.001	0.001
Truck 7.5 ÷ 12 [t]	0.149	[DB1], [DB2]	0.127	0.001	0.001
Truck 3.5 ÷ 7.5 [t]	0.165	[DB1], [DB2]	0.142	0.001	0.002
Truck 3.5 ÷ 7.5 [t][a]	0.156	[DB1], [DB2]	0.136	0.000	0.001
Truck <3.5 [t][b]	0.335	[DB1], [DB2]	0.291	0.001	0.001
Truck <3.5 [t][c]	0.554	[DB1]	0.442	0.003	0.005
Airplane	0.984	[DB1]	0.828	0.005	0.005

[a]Truck operated without append
[b]Light good vehicles
[c]Vans

Table 3.20 Classification factors adopted

	Impact categories		
Burdens	GW	AP	TOPP
CH_4	21		0.014
CO			0.11
CO_2	1		
H_2S		1	
HCl		0.88	
HF		1.6	
N_2O	320		
NH_3		1.88	
NMVOC			1
NO_x		0.7	1.22
SO_2		1	

The regression analysis on equivalent emissions data in Table 3.19 led to model the three unit mass of equivalent emissions e_i [kg/t·km] as a linear function of the loss factor:

$$e_i = \alpha_i \cdot f + \beta_i \quad i = (CO_2)_{eq.}, (SO_2)_{eq.}, (NMVOC)_{eq.} \,. \tag{3.33}$$

Regression parameters α_i and β_i and corresponding R^2 values, as well as the monetary cost per unit mass emission of the i-th pollutant ε_i as in [13] are in table 3.21.

Table 3.21 Regression parameters and unit monetary costs for the impact category considered

Impact	α_i	β_i	R^2	ε_i
i	[kg/t·km]	[kg/t·km]	[–]	[€/kg]
GW				
$(CO_2)_{eq.}$	0.8192	0.0094	0.9954	0.05
AP				
$(SO_2)_{eq.}$	0.0045	0.0001	0.9193	4.00
TOPP				
$NMVOC_{eq.}$	0.0051	0.0002	0.8443	30.00

Starting from data provided by an Italian logistic company (see Table 3.22) transport costs have been modeled here as:

$$\Phi_T = \frac{G}{Q} \cdot \left[z_1 \cdot e^{f \cdot z_2} \cdot m \cdot Q + T_1 \cdot e^{f \cdot T_2} \cdot L \cdot m \cdot Q \right] \tag{3.34}$$

In Eq. (3.33), the transport cost for a single order (φ_T, the terms in square bracket in Eq. 3.33) is obtained as the sum of the cost due to the number and the particular type of the Cargo Transport Unit (CTU), $\varphi_{T,CTU}$, and the cost related to the covered distance and the mass of products transported, $\varphi_{T,k}$:

$$\varphi_T = \varphi_{T,CTU} + \varphi_{T,k} \tag{3.35}$$

Both contributions have been shaped as a function of the loss factor value (f). In case of road transport, the CTU can be identified as the payload of the truck type considered. Results of a regression analysis carried out on the data provided by an Italian logistic company led to shape both contributions through exponential relationships (see Table 3.23):

$$\varphi_{T,CTU} = z_1 \cdot e^{f \cdot z_2} \cdot m \cdot Q \tag{3.36}$$

$$\varphi_{T,k} = T_1 \cdot e^{f \cdot T_2} \cdot L \cdot m \cdot Q \tag{3.37}$$

Table 3.22 Transport cost data adopted

Transport means	GVWR category	f	CTU/Payload [t]	$\varphi_{T,CTU}$ [€/CTU]	$\varphi_{T,CTU}$ [€/t]	$\varphi_{T,k}$ [€/t·km]
Ship					107.8	0.0032
Rail					88.0	0.0114
Truck	34 ÷ 40 [t]	0.0633	24.0	975	40.6	0.0219
Truck	26 ÷ 28 [t]	0.1193	12.0	731	60.9	0.0328
Truck	7.5 ÷ 12 [t]	0.1488	5.0	488	97.5	0.0525
Truck	3.5 ÷ 7.5 [t]	0.1651	3.0	423	140.8	0.0758
Airplane					40.3	0.2679

Table 3.23 Regression parameters values of transport costs functions

Transport means	z_1 [€/t]	z_2	T_1 [€/t·km]	T_2
Ship	107.800	0.000	0.032	0.000
Rail	88.000	0.000	0.011	0.000
Truck	17.780	11.740	0.010	11.740
Airplane	40.300	0.000	0.268	0.000

In case of maritime, rail and airfreight transport, data from [14] have been adopted (see Table 3.22). Parameters values adopted in (3.36) and (3.37) are in Table 3.23.

By adopting (3.25), (3.31), (3.32), and (3.34), the logistic cost function (3.19) can be rewritten as:

$$\begin{aligned}\Phi_L = {} & c_h \cdot \left[\frac{1}{2} \cdot \frac{G \cdot LT}{H \cdot p} \cdot + LT \cdot (D^* - E(D)) \right] + G \cdot m \cdot \left[z_1 \cdot e^{f \cdot z_2} + (T_1 \cdot e^{f \cdot T_2}) \cdot L \right] + \\ & + c_s \cdot \frac{p \cdot H}{(LT)^{3/2} \cdot \sigma_D \cdot \sqrt{2\pi}} \cdot \int_{D^*_{TOT}}^{+\infty} \exp\left[- \frac{(D_{TOT} - LT \cdot E(D))^2}{2 \cdot LT \cdot \sigma_D^2} \right] \cdot \left(D_{TOT} - D^*_{TOT} \right) dD_{TOT} + \\ & + G \cdot m \cdot L \cdot \left[f \cdot \sum_i \varepsilon_i \cdot \alpha_i + \sum_i \varepsilon_i \cdot \beta_i \right]\end{aligned} \tag{3.38}$$

And the corresponding logistic cost factor as:

$$F_L = \left[\frac{LT}{H \cdot p} \cdot + \frac{LT \cdot (D^* - E(D))}{2 \cdot G} \right] + 2 \cdot c_h \cdot m \cdot \left[z_1 \cdot e^{f \cdot z_2} + (T_1 \cdot e^{f \cdot T_2}) \cdot L \right] + \tag{3.39}$$

$$+ \frac{2}{G} \cdot \frac{c_S}{c_h} \cdot \frac{p \cdot H}{(LT)^{3/2} \cdot \sigma_D \cdot \sqrt{2\pi}} \cdot \int_{D^*_{TOT}}^{+\infty} \exp\left[-\frac{(D_{TOT} - LT \cdot E(D))^2}{2 \cdot LT \cdot \sigma_D^2} \right] \cdot \left(D_{TOT} - D^*_{TOT} \right) dD_{TOT} +$$

$$+ \frac{2}{c_h} \cdot m \cdot L \cdot \left[f \cdot \sum_i \varepsilon_i \cdot \alpha_i + \sum_i \varepsilon_i \cdot \beta_i \right]$$

The general logistic optimization problem (3.29) has been solved for different value of *SL* (90 %–95 %–99 %) and *p* (0.1 ÷ 0.9) in order to evaluate the effect of the demand variability. Different values of the distance L (200–400–1000 [km]) have been considered. The upper limit value for the parameter p, due to organizational constraints, has been assumed equal to 0.9 instead of 1.0. Results obtained are in Table 3.24.

Table 3.24 Results obtained in case of $c_S/c_h = 0.65$

$L = 200$ [km]					
SL	f_{OPT}	p_{OPT}	F_L	*SOQ* [unit]	*SS* [unit]
0.90	0.063	0.4	0.0161	24.72	1.67
0.95	0.063	0.5	0.0142	19.77	2.14
0.99	0.063	0.9	0.0122	10.99	3.04
$L = 400$ [km]					
SL	f_{OPT}	p_{OPT}	F_L	*SOQ* [unit]	*SS* [unit]
0.90	0.063	0.6	0.0191	32.96	2.35
0.95	0.063	0.9	0.0169	21.97	3.03
0.99	0.063	0.9	0.0152	21.97	4.29
$L = 1000$ [km]					
SL	f_{OPT}	p_{OPT}	F_L	*SOQ* [unit]	*SS* [unit]
0.90	0.063	0.9	0.0269	54.93	3.70
0.95	0.063	0.9	0.0252	54.93	4.77
0.99	0.063	0.9	0.0244	54.93	6.77

The transport means with the highest capacity (the slowest ones) minimize logistical and environmental costs for each distance and service level considered even if the environmental costs are significantly lower than transport costs (10 %).

For each distance considered L, as expected, the safety stock size (SS) increases with the increase of the service level (SL) fulfilled While for each SL value the SS size increases with L.

For each distance considered, the service level (SL) differently influences the optimal order quantity. The optimal order quantity values (SOQ) decrease, with the increase of the service level, for low distances ($L \leq 400$ [km]); in fact, in case of high SL, smaller order quantities lead to an increase in shortage costs (see Eq. 3.22) lower than the corresponding decrease of the holding costs.

The increase of the service level achieved does not affect the solution of the problem in terms of order quantity for high distances ($L = 1000$ [km]). In this case, in fact, holding costs are significantly higher than shortage costs: the minimum of the logistic cost function is therefore obtained for the lowest values of the order quantity adoptable (when $p = 0.9$) compliant with lead time constraints.

A sensitivity analysis has been carried out varying the unit shortage cost c_S ($c_S = \tau \cdot c_h$, with $\tau = 1, 2, 10$) in order to evaluate the effect of the shortage costs on the solution of problem (3.28).

The case of a unit shortage cost greater than the unit holding cost refers to low value-added products and high unit profit, as in case of mass production of high technology product.

Results obtained in case of a distance equal to 200 [km] are in Table 3.25.

In case of small transport distance, the solution of the problem is not affected by the unit shortage costs in terms of transport means selection since the optimal loss factor values does not change and is characteristic of the slowest and high-capacity means of transport. On the contrary, with the increase of the unit shortage costs, higher optimal order quantities are obtained as solution of the logistics problem. The increase of the transport distances ($L = 400$ [km], 1000 [km]) leads to the same remarks: the effect of increasing unit shortage cost on the solution of (3.29) is similar to the one observed in case of a short transport distances. In this case the increase in the optimal order quantity with the increase of unit shortage cost is less severe if compared with the short distance case due to the higher transport cost.

Table 3.25 Results of the c_S/c_h sensitivity analysis in case of L = 200 [km]

		SL = 90 %			
				SOQ	SS
c_s/c_h	f_{OPT}	p_{OPT}	F_L	[unit]	[unit]
0.65	0.063	0.4	0.0161	24.72	1.67
1.00	0.063	0.3	0.0175	32.96	1.67
2.00	0.063	0.2	0.0206	49.43	1.67
10.00	0.063	0.1	0.0336	98.87	1.67
		SL = 95 %			
				SOQ	SS
c_s/c_h	f_{OPT}	p_{OPT}	F_L	[unit]	[unit]
0.65	0.063	0.5	0.0142	19.77	2.14
1.00	0.063	0.4	0.0152	24.72	2.14
2.00	0.063	0.3	0.0172	32.96	2.14
10.00	0.063	0.1	0.0266	98.87	2.14
		SL = 99 %			
				SOQ	SS
c_s/c_h	f_{OPT}	p_{OPT}	F_L	[unit]	[unit]
0.65	0.063	0.9	0.0122	10.99	3.04
1.00	0.063	0.9	0.0125	10.99	3.04
2.00	0.063	0.8	0.0133	12.36	3.04
10.00	0.063	0.3	0.0168	32.96	3.04

The effects of the demand stochastic variability on the solution of Eq. (3.29) have been evaluated through a sensitivity analysis. The logistics problem has been solved in case of a cv values of 0.5 and 0.8, c_S/c_h = 0.65 and L = 200 [km] (Fig. 3.10 and Fig. 3.11).

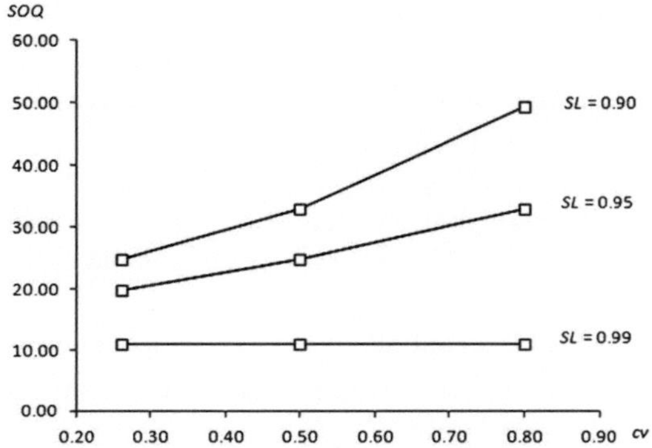

Fig. 3.10 SOQ vs. cv values in case of L = 200 [km] and c_S/c_h = 0.65

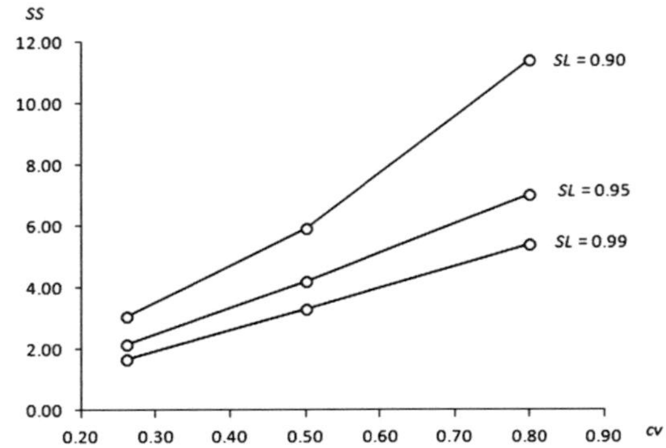

Fig. 3.11 *SS* vs. *cv* values in case of $L = 200$ [km] and $c_S/c_h = 0.65$

Again, for each service level (*SL*) considered and even in case high *cv* values, the slowest transport means minimize the logistic cost function. As expected, the increase in the demand variability leads to an increase of the *SS* level required. The demand variability also affects the optimal order quantity values. The increase in the optimal SOQ, however, reveals to be dependent to the service level fulfilled: in case of lower *SL* values, a greater increase of the *SOQ* values is observed.

3.2.2 Lead Time Uncertainty

The SOQ model is here defined in case of stochastic variability of supply lead time, under the following assumptions:

a. the product demand (*D*) is deterministic and stationary, $D = G/H$;
b. orders do not cross;
c. backordering is not allowed;
d. the expected value of the supply lead time ($E(LT)$) is evaluated as the sum of the expected transport time ($E(T_T)$) and the expected value of the time required for the material handling, order management and quality control ($E(T_L)$); $E(T_L)$ is evaluated as a fraction ($k < 1$) of $E(T_T)$; consequently, speed of transport is obtained from the free flow or cruise speed (v_{act}) of different transport means as $v = v_{act}/(1 + k)$, and expected value of the supply lead time as $E(LT) = L/v$;
e. supply lead time (*LT*) is a continuous random variable normally distributed.

The logistic cost function considered in case of stochastic variability of supply lead time is:

$$\Phi_L = \Phi_O + \Phi_H + \Phi_T + \Phi_S + \Phi_{EX} \quad [€/\text{year}] \tag{3.40}$$

As already discussed in Sect. 2, in case of stochastic variability of the supply lead time (LT), three different situations may occur:

(D) $LT \leq E(LT)$
(E) $E(LT) < LT \leq LT^*$
(F) $LT > LT^*$

with LT^* the maximum value of the supply lead time not causing a stock out event at a given service level (SL), and

$$SL = prob\left(D_{TOT} \leq LT^* \cdot D\right) = \int_{-\infty}^{LT^*} pdf(LT)dLT = F(LT^*) \tag{3.41}$$

As a consequence, in order to evaluate the holding costs in Eq. (3.40) the expected inventory level ($E(I)$) in one ordering cycle has to be computed. With $E(I)_A$, $E(I)_B$, and $E(I)_C$ the expected inventory level in the three cases (Table 3.26), the expected inventory level during one ordering cycle can be evaluated as:

$$\begin{aligned} E(I) = & \int_{-\infty}^{E(LT)} E(I)_A \cdot pdf(LT)dLT \\ & + \int_{E(LT)}^{LT^*} E(I)_B \cdot pdf(LT)dLT + \int_{LT^*}^{+\infty} E(I)_C \cdot pdf(LT)dLT \cdot \end{aligned} \tag{3.42}$$

Table 3.26 Expected inventory level and ordering cycle length in the three cases considered

	Expected inventory level E(I)	
Case A $LT \leq E(LT)$	$E(I)_A = \frac{1}{2}Q + D \cdot [E(LT) - LT] + SS$	
Case B $E(LT) < LT \leq LT^*$	$E(I)_B = \frac{1}{2}Q - D \cdot [LT - E(LT)] + SS$	
Case C $LT > LT^*$	$E(I)_C = \frac{1}{2}Q + \frac{1}{2}SS - \frac{1}{2}Q \cdot \frac{[LT^* - E(LT)]}{CT_C}$	
	Consumption time CT	*Case probability*
Case A $LT \leq E(LT)$	$\frac{Q}{D}$	$\int_{-\infty}^{0} pdf(z)dz$
Case B $E(LT) < LT \leq LT^*$	$\frac{Q}{D}$	$\int_{0}^{z^*} pdf(z)dz$
Case C $LT > LT^*$	$\frac{Q}{D} + [LT - LT^*]$	$\int_{z^*}^{+\infty} pdf(z)dz = 1 - SL$

Under the assumption of $E(I)_C = ½\cdot Q$ (see Sect. 2), the expected inventory level $E(I)$ in one ordering cycle can be computed as:

$$E(I) = \frac{1}{2}Q + D \cdot \sigma_{LT} \cdot [pdf(z^*) + z^* \cdot F(z^*)], \tag{3.43}$$

and the corresponding holding cost is:

$$\Phi_H = c_H \cdot \left\{ \frac{1}{2}Q + D \cdot \sigma_{LT} \cdot [pdf(z^*) + z^* \cdot F(z^*)] \right\}. \tag{3.44}$$

In Eqs. (3.43) and (3.44), z is the standardized variable of the stochastic variable LT, and hence:

$$z^* = \frac{LT^* - E(LT)}{\sigma_{LT}}. \tag{3.45}$$

As far as concern shortage costs, they occur only in case C:

$$\Phi_S = c_S \cdot \frac{G}{Q} \cdot D \cdot \sigma_{LT} \cdot L(z^*), \tag{3.46}$$

with

$$L(z^*) = \int_{z^*}^{+\infty} (z - z^*) \cdot pdf(z)\, dz = pdf(z^*) - z^* \cdot [1 - F(z^*)]. \tag{3.47}$$

In the model, quadratic functions are adopted in order to shape the dependency of both the transport cost and the external cost of transport on loss factor values:

$$\Phi_T(f,L) = G \cdot m \cdot [a \cdot f^2 + b \cdot f + c]. \tag{3.48}$$

$$\Phi_{EX} = G \cdot m \cdot L \cdot [\delta \cdot f^2 + \mu \cdot f]. \tag{3.49}$$

The logistics cost function can be finally reformulated as:

$$\Phi_L = c_H \cdot \left\{ \frac{1}{2} Q + D \cdot \sigma_{LT} \cdot \left[pdf(z^*) + z^* \cdot F(z^*) \right] \right\} + c_O \cdot \frac{G}{Q} + \left[a \cdot f^2 + b \cdot f + c \right] \cdot G \cdot m$$

$$+ c_S \cdot \frac{G}{Q} \cdot D \cdot \sigma_{LT} \cdot L(z^*) + G \cdot m \cdot L \cdot \left[\delta \cdot f^2 + \mu \cdot f \right], \tag{3.50}$$

and the corresponding logistic costs factor as

$$F_L = \frac{2}{G} \cdot \left\{ \frac{1}{2} Q + D \cdot \sigma_{LT} \cdot \left[pdf(z^*) + z^* \cdot F(z^*) \right] \right\} + \frac{c_O}{c_h} \cdot \frac{2}{Q} + \frac{2}{c_h} \cdot m \cdot \left[a \cdot f^2 + b \cdot f + c \right]$$

$$+ 2 \cdot \frac{c_S}{c_h} \cdot \frac{D}{Q} \cdot \sigma_{LT} \cdot L(z^*) + \frac{2}{c_h} \cdot m \cdot L \cdot \left[\delta \cdot f^2 + \mu \cdot f \right]. \tag{3.51}$$

For an assigned set of (SL, cv_{LT}, L) values, with $\sigma_{LT} = cv_{LT} \cdot E(LT)$, solving problem (3.29) the optimal values of loss factor (f_{OPT}), order quantity (SOQ), reorder level ($r(f_{OPT})$), and of the safety stock ($SS(f_{OPT})$) are obtained.

The following procedure has been adopted here to solve problem (3.26):

i. compute LT^* from (3.41);
ii. compute z^* from (3.45);
iii. compute f_{OPT} and p_{OPT} by solving (3.29);
iv. compute $E(LT) = L/v(f_{OPT})$;
v. compute $r(f_{OPT}) = D \cdot E(LT)$;
vi. compute $SOQ = r(f_{OPT}) \cdot p_{OPT}$;
vii. compute $\sigma_{LT} = cv_{LT} \cdot E(LT)$;
viii. compute $SS(f_{OPT})$ by means of:

$$SS = D \cdot z^* \cdot \sigma_{LT} \tag{3.52}$$

Step (iii) has to be carried out by means of a numerical method.

Case Study

The model has been applied to the case study discussed in Sect. 3.2.1 [9, 10]. Following the procedure above, the logistics problem (3.29) has been solved for PX product in site 1 ($G = 9224$ [unit/year], $H = 3520$ [h/year], $D = 2.62$ [unit/h]). Cost data in [€$_{2013}$] have been considered ($c_H = 18.98$ [€$_{2013}$/unit·year], $c_S = 12.32$ [€$_{2013}$/unit]). A unit order cost (c_O) of 100 [€/order] has been assumed. The corresponding EOQ value is 312 [unit/order].

Loss factor and average speed of transport values as in Sect. 3.2.1 and unit external cost data of water, rail, road, and air freight transport as in Sect. 3.1.2 ([€$_{2013}$/t·km]) have been adopted.

Parameter values adopted to evaluate unitary transport cost are in Table 3.27.

Table 3.27 Parameters values for unitary transport cost evaluation per transport distance

L km	*a* [€/kg]	*b* [€/kg]	*c* [€/kg]
200	391.37	–402.35	108.69
500	329.84	–259.17	107.63
1000	227.28	–20.54	105.87

EOQ and SOQ Comparison (In Case of Deterministic Supply Lead Time)

The hypothesis of deterministic supply lead time ($cv = 0$) allows to perform a preliminary comparison between results of the *SOQ* model proposed with those obtained by the traditional *EOQ* model [15].

The results obtained without considering cost of externalities ("*econ.*") are compared with results obtained considering external costs ("*sust.*") (Table 3.28).

In the sustainable solution, a *SOQ* value greater than the *EOQ* value is identified as optimal choice for each transport distance considered. When the transport distance considered increases, the optimal loss factor values decrease both in case of economic and in case of sustainable solutions. As a consequence, in case of long transport distance higher reorder levels are required. For the sustainable solution, when compared with the economic one, slower transport means (and higher reorder levels) are identified as optimal choice.

Finally, as expected, sustainable solutions are characterized by higher logistics cost function (F_L) values than the corresponding ones in the economic case.

Table 3.28 Sustainable and economic solutions comparison in case of $cv = 0$

L km	*f sust.*	*SOQ sust.* unit	F_L *sust.*	*r sust.* unit
200	0.149	318	0.071	12
500	0.037	318	0.073	27
1000	0.037	330	0.074	55
L km	*f econ.*	*SOQ econ.* unit	F_L *econ.*	*r econ.* unit
200	0.335	308	0.069	8
500	0.335	308	0.071	20
1000	0.037	318	0.073	55

Evaluating the Effects of the Supply Lead Time Variability

In case of a variable supply lead time, shortage costs affect solution of (3.29). Shortage costs depend on safety stock level (*SS*), which is a function of the lead time variability and of the *SL* (see Eq. 3.52). In order to investigate the magnitude of the effects of the supply lead time variability on the solutions of the logistics problem (3.29), it has been solved for different values of the lead time variability (*cv*) and of the service level (*SL*). A negligible influence of the service level values on the solution of the logistics problem (f_{OPT}, *SOQ*, $r(f_{OPT})$) has been observed. This is due to the small value of the unitary shortage cost (c_S) considered.

As an example, results obtained in case of $SL = 0.95$ are in Table 3.29 and Table 3.30. Road transport is identified as optimal choice for each level of supply lead time variability considered in case of short transport distance ($L = 200$ [km]). Fast road transport means are solution of problem (3.29) when a high variability of the lead time is considered. They are solution of (3.29) for higher supply lead time variability value in case of long transport distance ($L = 500$–1000 [km]). Transport means characterized by better environmental performances (rail, ship) are the optimal choice in the other cases considered (see Table 3.29). Reorder level obtained ($r(f_{OPT})$, see Table 3.30) are consistent with the loss factor values identified as optimal by means of (3.29): smaller reorder level can be adopted in case of faster transport means. In case of short transport distance and low variability of supply lead time, SOQ values higher than EOQ are observed. Furthermore, supply lead time variability weakly affects SOQ values observed. On the contrary, an increase in the supply lead time variability (and in the transport distance) leads to higher *SS* values (see Table 3.30). The effect of the supply lead time variability on total logistic costs ($F_L(f_{OPT.})$, see Table 3.30) proved to be dependent on the transport distance considered. In the two limit cases considered ($cv = 0.00$ and $cv = 2.00$, $SL = 0.95$) an increase in total annual costs of 5.5 %, 18.7 %, and 45.7 % is observed in case of a transport distance of 200 [km], 500 [km], and 1000 [km], respectively (see Table 3.30).

Table 3.29 Optimal means of transport (f_{OPT}) and *SOQ* values for different *L* and *cv* values in case of $SL = 0.95$

	f_{OPT}				*SOQ*		
	L [km]				*L* [km]		
cv	200	500	1000	*cv*	200	500	1000
0.00	*Truck 7.5–12 [t]*	*Rail—electric*	*Rail—electric*	0.00	*318*	*318*	*330*
0.10	*Truck 7.5–12 [t]*	*Rail—electric*	*Ship*	0.10	*318*	*318*	*318*
0.25	*Vans*	*Rail—electric*	*Ship*	0.25	*319*	*318*	*318*
0.50	*Vans*	*Rail—electric*	*Ship*	0.50	*319*	*318*	*318*
0.75	*Vans*	*Vans*	*Ship*	0.75	*319*	*319*	*318*
1.00	*Vans*	*Vans*	*Ship*	1.00	*319*	*319*	*318*
2.00	*Vans*	*Vans*	*Vans*	2.00	*319*	*319*	*318*

Table 3.30 Reorder level ($r(f_{OPT})$), SS, and F_L values for different L and cv values in case of $SL = 0.95$

	$r(f_{OPT})$		
	L [km]		
cv	200	500	1000
0.00	*25*	*55*	*109*
0.10	*25*	*55*	*102*
0.25	*7*	*55*	*102*
0.50	*7*	*55*	*102*
0.75	*7*	*19*	*102*
1.00	*7*	*19*	*102*
2.00	*7*	*19*	*37*

	$SS(f_{OPT})$		
	L [km]		
cv	200	500	1000
0.00	*0*	*0*	*0*
0.10	*2*	*4*	*8*
0.25	*2*	*11*	*21*
0.50	*3*	*22*	*42*
0.75	*5*	*12*	*63*
1.00	*6*	*15*	*84*
2.00	*12*	*31*	*61*

	$F_L(f_{OPT}, p_{OPT})$		
	L [km]		
cv	200	500	1000
0.00	*0.070*	*0.073*	*0.074*
0.10	*0.071*	*0.074*	*0.076*
0.25	*0.071*	*0.076*	*0.080*
0.50	*0.072*	*0.079*	*0.085*
0.75	*0.072*	*0.081*	*0.091*
1.00	*0.073*	*0.082*	*0.097*
2.00	*0.074*	*0.087*	*0.107*

In order to investigate the effect of the external costs on the solution of problem (3.29), it has been solved in case no external costs are computed in the logistics cost function ($\Phi_{EX} = 0$). Results obtained in case of a service level (SL) equal to 0.95 are in Table 3.31.

Table 3.31 Optimal loss factor (f_{OPT}) values of the sustainable and of the economic ($\Phi_{EX} = 0$) solution

	f_{OPT} sust.		
	L [km]		
cv	200	500	1000
0.00	*0.149*	*0.037*	*0.037*
0.10	*0.149*	*0.037*	*0.010*
0.25	*0.554*	*0.037*	*0.010*
0.50	*0.554*	*0.037*	*0.010*
0.75	*0.554*	*0.554*	*0.010*
1.00	*0.554*	*0.554*	*0.010*
2.00	*0.554*	*0.554*	*0.554*

	f_{OPT} econ.		
	L [km]		
cv	200	500	1000
0.00	*0.335*	*0.335*	*0.037*
0.10	*0.335*	*0.335*	*0.010*
0.25	*0.554*	*0.554*	*0.554*
0.50	*0.554*	*0.554*	*0.554*
0.75	*0.554*	*0.554*	*0.554*
1.00	*0.554*	*0.554*	*0.554*
2.00	*0.554*	*0.554*	*0.984*

As showed in Table 3.31, the internalization of the external costs leads to identify as optimal choice slower transport means (3.29). As far as concern SOQ values, negligible differences have been observed in the solutions obtained in the two cases.

A sensitive analysis has been carried in order to evaluate the effects of a change in the ordering cost on the solutions of (3.29). For this purpose, problem (3.29) has been solved considering other two unitary order cost values (10 [€/order] and 200 [€/order]). Results obtained are in Table 3.32. Solutions obtained proved to be slightly affected by order cost values, except for *SOQ* values: with the increase of fixed order cost, higher values of *SOQ* minimize total logistic costs (see Table 3.32 and Table 3.29).

Table 3.32 *SOQ* values for different L, cv, and c_O values in case of $SL = 0.95$

SOQ	c_O = 10 [€/order]			c_O = 200 [€/order]		
	L [km]			L [km]		
cv	200	500	1000	200	500	1000
0.00	*93*	*98*	*100*	*403*	*420*	*420*
0.10	*93*	*99*	*100*	*403*	*420*	*420*
0.25	*93*	*99*	*98*	*403*	*420*	*464*
0.50	*93*	*98*	*109*	*374*	*420*	*464*
0.75	*93*	*98*	*109*	*374*	*467*	*464*
1.00	*93*	*104*	*104*	*374*	*467*	*464*
2.00	*93*	*104*	*98*	*374*	*467*	*467*

3.2.3 SOQ of Repairable Spare Parts with Uncertain Demand

In this Section, the SOQ model is formulated in order to solve the single-product replenishment problem in case of a reparable spare parts inventory and a stochastic variability of the spare parts demand. In the logistic cost function both economic and environmental costs relating to repair and replacement of spare parts are considered. Environmental costs of the production of new parts and of the disposal of used ones are taken into account. The logistic cost function has been modified here in order to consider the effects on the logistics costs of the repair policy adopted. Repair policies considered differ for the percentage of spare parts repaired and not replaced. Results of a case study from the automotive components industry are presented and discussed.

Although the SOQ model adopted here is similar to the one in Sect. 3.2.1, further notations (see Table 3.33) and assumptions (listed below) have been adopted in this Section.

Table 3.33 Further notations adopted in the SOQ model of reparable spare parts

Variable	Name	Unit
Φ_{PR}	Purchase and repair costs	[€/year]
χ	Spare parts repair rate	[–]
ψ	Spare parts repair success rate	[–]
G_R	Successfully repaired spare parts	[unit/year]
G_N	Quantity of new ones to be purchased	[unit/year]
Q_R	Spare parts successfully repaired per cycle	[unit/cycle]
Q_N	New spare parts purchased per cycle	[unit/cycle]
c_N	Unit purchase cost	[€/unit]
c_R	Unit repair cost	[€/unit]

The following assumptions are assumed in defining the SOQ model in case of stochastic variability of spare parts demand:

1. the expected annual demand (G) is known;
2. the product demands in each time period, D_i, are independent stochastic variables;
3. the supply lead time is evaluated as $LT = L/v$, with v the average speed in transport (see Sect. 3.2.1);
4. in each i-th period of the lead time (LT), product demand is characterized by the same expected value and by the same standard deviation:

$$E(D_1) = E(D_2) = \ldots = E(D_{LT}) = E(D) = G/H, \tag{3.53}$$

$$\sigma_{D_1}^2 = \sigma_{D_2}^2 = \ldots = \sigma_{DLT}^2 = \sigma_D^2; \tag{3.54}$$

5. recovering processes are performed in house;
6. the success rate in repairing activities (ψ) is deterministically known; it ranges in [0; 1];
7. repaired spare parts are stocked in the inventory;
8. the amount of annual replaced items replaced cannot be null, since items cannot be repaired forever, they are subjected to obsolescence, and recovering processes are characterized by a success rate (χ) less than 100 %; χ values in the range [0; χ_{max}], with $\chi_{max} < 1$, have been considered.

In the case of repairable spare parts, the annual logistics cost function can be rewritten as:

$$\Phi_L = \Phi_{PR} + \Phi_O + \Phi_H + \Phi_T + \Phi_S + \Phi_{EX}. \tag{3.55}$$

The annual logistics costs depend on the inventory level at the beginning of each ordering cycle (Q), on the repair rate (χ), as well as on the transport means adopted (f):

$$\Phi_L = \Phi_L(f, \chi, Q), \tag{3.56}$$

and the general logistics optimization problem is defined here as:

$$\min_{f,\chi,Q} \Phi_L \tag{3.57}$$

In Eq. (3.55), purchase and repair costs of spare parts are explicitly considered, since the repair policy adopted (percentage of repaired spare parts) affects the overall logistics costs.
Cost figures in Eq. (3.55) under the hypothesis of repairable spare parts are detailed in the following.

<u>*Purchase and Repair Costs*</u>

The expected value of the annual demand of spare parts (G) is the sum of the spare parts successfully repaired (G_R) and of the purchased spare parts (G_N):

$$G_R = \chi \cdot \psi \cdot G, \tag{3.58}$$

$$G_N = (1 - \chi \cdot \psi) \cdot G. \tag{3.59}$$

The total annual costs of purchasing and repairing the spare parts is:

$$\Phi_{PR} = c_N \cdot G_N + c_R \cdot \chi \cdot G. \tag{3.60}$$

Unsuccessfully repaired process of a spare part generates an extra cost since the cost of repairing it will be added to the purchase cost of replacing it.

Holding Costs

At the beginning of each ordering cycle, the quantity (Q) of spare parts in the inventory consists of new spare parts purchased per cycle (Q_N), and of successfully repaired spare parts per cycle (Q_R) (see Fig. 3.12):

$$Q_N = (1 - \chi \cdot \psi) \cdot Q, \tag{3.61}$$

$$Q_R = \chi \cdot \psi \cdot Q. \tag{3.62}$$

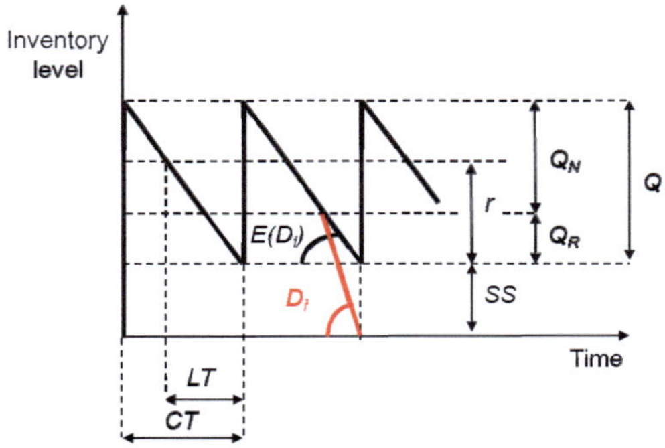

Fig. 3.12 Spare parts inventory level over time

With *SS* the safety stock level adopted, the corresponding annual holding costs can be evaluated as:

$$\Phi_H = c_H \left[(1 + \chi \cdot \psi) \cdot \frac{Q}{2} + SS \right]. \tag{3.63}$$

By considering (3.3) and (3.23), Eq. (3.63) can be expressed as:

$$\Phi_H = \frac{c_H \cdot L}{v} \left[(1 + \chi \cdot \psi) \cdot \frac{G \cdot L}{2 \cdot p \cdot H} + D^* - E(D) \right]. \tag{3.64}$$

Ordering Costs

With N_P the number of annual orders of spare parts:

$$N_P = \frac{G_N}{Q_N} = \frac{G}{Q}, \tag{3.65}$$

the annual ordering costs are obtained as:

$$\Phi_O = c_O \frac{G}{Q} = c_O \frac{p \cdot H \cdot v}{L}. \tag{3.66}$$

Transport Cost

The total annual costs of transport are computed as:

$$\Phi_T = c_T(f, L) \cdot (1 - \chi \cdot \psi) \cdot G \cdot m. \tag{3.67}$$

Shortage Costs

The yearly shortage costs are computed as:

$$\Phi_S = c_S \frac{G}{Q} N_S, \tag{3.68}$$

and by considering (3.25):

$$\Phi_S = c_S \cdot \frac{G}{Q} \cdot \int_{D^*_{TOT}}^{+\infty} pdf(D_{TOT}) \cdot (D_{TOT} - D^*_{TOT}) dD_{TOT}, \tag{3.69}$$

where D^*_{TOT} is the maximum lead time demand not causing a stock-out for a given service level (SL):

$$SL = prob\left\{D_{TOT} \le LT \cdot D^*\right\} = \int_{-\infty}^{LT \cdot D^*} pdf(D_{TOT}) dD_{TOT}. \tag{3.70}$$

Environmental Cost

Environmental costs are evaluated here as the sum of the environmental cost of spare parts replacing and the environmental cost of new spare parts transport. With c_E [€/unit] the environmental cost of replacing one item (due to both the disposal of the replaced item and the production of the new one), the environmental costs can be computed as:

$$\Phi_E = (1-\chi\cdot\psi)\cdot G\cdot\left[c_E + m\cdot L\cdot\sum_j \varepsilon_j e_j(f)\right] \quad (3.71)$$

The Optimization Problem

By introducing the cost figures in Eqs. (3.60), (3.64), (3.66), (3.67), (3.69), and (3.71) in Eq. (3.55), the logistic costs function can be expresses as:

$$\begin{aligned}\Phi_L &= c_N\cdot G_N + c_R\cdot\chi\cdot G + \frac{c_H\cdot L}{v}\left[(1+\chi\cdot\psi)\cdot\frac{G\cdot L}{2\cdot p\cdot H} + D^* - E(D)\right] + \\ &+ c_O\frac{p\cdot H\cdot v}{L} + c_T(f,L)\cdot(1-\chi\cdot\psi)\cdot G\cdot m \\ &+ c_S\cdot\frac{G}{Q}\cdot\int_{D^*_{TOT}}^{+\infty} pdf(D_{TOT})\cdot(D_{TOT}-D^*_{TOT})dD_{TOT} + (1-\chi\cdot\psi)\cdot G\cdot\left[c_E + m\cdot L\cdot\sum_j \varepsilon_j e_j(f)\right].\end{aligned} \quad (3.72)$$

The logistics problem can be defined by using the logistic cost factor F_L as:

$$\min_{f,p,\chi} F_L \quad (3.73)$$

By solving problem (3.73), the optimal means of transport (f_{OPT}), the optimal repair rate (χ_{OPT}), the sustainable order quantity $SOQ(f_{OPT}, \chi_{OPT}, p_{OPT})$ (see Eq. 3.74), and the corresponding optimal safety stock level $SS(f_{OPT})$ (see Eq. 3.75) are obtained.

$$SOQ = \frac{(1-\chi_{OPT}\cdot\psi)\cdot G\cdot L}{p_{OPT}\cdot H\cdot v(f_{OPT})} \quad (3.74)$$

$$SS = \frac{L}{v(f_{OPT})}\left[D^*(f_{OPT}) - E(D)\right] \quad (3.75)$$

Numerical Experiments

Results of the application of the model to a full scale case study from the automotive components industry in case of spare parts demand varying stochastically [16] are in the following presented and discussed.

Solutions of problem (3.73) have been obtained for different values of the transport distance (L), of the service level (SL), of the coefficient of variation of the demand (cv), and of the unit repair cost (c_R).

Parameter values adopted in the numerical experiments are from [17], [18], and [19], are summarized in Table 3.34.

In order to evaluate the effect of the environmental costs on the solutions of the logistics problem, a preliminary comparison has been carried out with the traditional EOQ model [15]. In Table 3.35, solutions obtained from the EOQ model of Harris are compared with those obtained from (3.73) for different transport distance (L) and demand variability (cv), in case of $\chi = 0$ and $SL = 0.9$.

Table 3.34 Parameters values adopted

Parameter	Value	Unit
G	10,000	unit/year
χ	0.6; 0.9	–
cv	0.0; 0.1; 0.3	–
SL	0.90; 0.95; 0.99	–
H	3500	h/year
L	200; 500; 1000	km
m	25	kg/unit
c_N	40	€/unit
c_R	32; 40; 48	€/unit
c_H	500	€/unit
c_O	200	€/order
c_S	1000	€/unit
c_E	10	€/unit

Table 3.35 EOQ and SOQ model results comparison

	L [km]								
	200			500			1000		
cv	0.0	0.1	0.5	0.0	0.1	0.5	0.0	0.1	0.5
EOQ [unit]	89	89	89	89	89	89	89	89	89
SOQ [unit]	105	105	105	90	92	115	92	103	132
f_{OPT}	0.475	0.475	0.475	0.134	0.010	0.010	0.010	0.010	0.010
$F_L(EOQ)$	0.178	0.178	0.178	0.186	0.188	0.188	0.188	0.188	0.188
$F_L(SOQ)$	0.221	0.222	0.227	0.228	0.231	0.237	0.229	0.232	0.240
SS [unit]	–	1	4	–	1	5	–	1	7

In the case of a demand assumed as constant ($cv = 0.0$), the SOQ values obtained are higher than the corresponding EOQ values, since internal and external transport costs lead the SOQ model towards solutions characterized by higher order quantity values.

At the same time, the effect of demand variability on the solution provided by the SOQ model is of different magnitude depending on the transport distance (L) considered. Demand variability does not affect the solution in terms of both transport means selection (f value is constant) and optimal order quantity in the case of low transport distance. Optimal order quantity values increase with the demand variability in the case of higher transport distances ($L \geq 500$ [km]). Moreover, the increase of the SOQ is maximum for the longest distance considered ($L = 1000$ [km]). When a higher demand variability (cv) is considered, an increase in the SOQ values is observed. This is due to the influence of the shortage costs. For an assigned service level (SL), in fact, an increase of the demand variability (cv) leads to an increase in the number of stock out events in each ordering cycle (see Sect. 4.1). The annual shortage costs depend on the overall number of stock out events in the year, which in turn depends by both the number of stock out events per cycle and the number of the overall ordering cycles in the year. In case of high variability of the spare parts demand, the model identifies as optimal high SOQ values, since this leads to a reduction of the overall number of ordering cycle.

The magnitude of the effect of the demand variability on the SOQ values is not the same for all the transport distances considered. In case of short transport distance (L = 200 [km]), demand variability does not affect SOQ values. This could be easily explained by considering the optimal loss factor values identified by the model and listed in Table 3.36: in case of short transport distance, fast transport means are identified as optimal choice. As a consequence, small lead time (LT) values are obtained, and the effect of the demand variability on SOQ values becomes negligible.

The EOQ model does not consider the transport selection problem. For this reason, with the purpose to compare annual logistics costs of the solutions obtained by both the EOQ and the SOQ model, they have been computed, in case of the EOQ model solutions, by adopting the same transport means identified as optimal by the SOQ model. Total annual logistics costs of the SOQ model solutions are greater than the corresponding costs of the EOQ model solutions in all cases considered, because of the external costs considered in the first model. The differences observed increase with the increase in the demand variability due to the increase of the holding costs (a higher SS level is required) and the shortage costs.

Problem (3.73) has been solved for each SL, cv, c_R, and χ values considered. As an example, the trends of the logistics cost factor (F_L) vs. the loss factor (f) for the three transport distance (L) values considered in case of $\chi = 0.5$, $\psi = 0.9$, $SL = 0.95$, $c_v = 0.1$, $c_R = c_N$, and $p = 0.5$ are depicted in Fig. 3.13. As expected, the model identifies slow transport means as optimal choice when a long transport distance is considered, due to their good environmental performances.

Results obtained from numerical experiments showed that the repair success rate (ψ) affects solution of (3.39) only in terms of total annual logistics costs in case of a deterministic spare parts demand, and that in case of an uncertain demand, the repair option is cost-effective only in case of a high repair success rate performed. Results obtained in case of a repair success rate (ψ) of 0.9 and a unit repair cost equal to the unit purchase cost ($c_R = c_N$) for different transport distance (L), demand variability level (cv), and service level (SL) are in Tables 3.36–3.38.

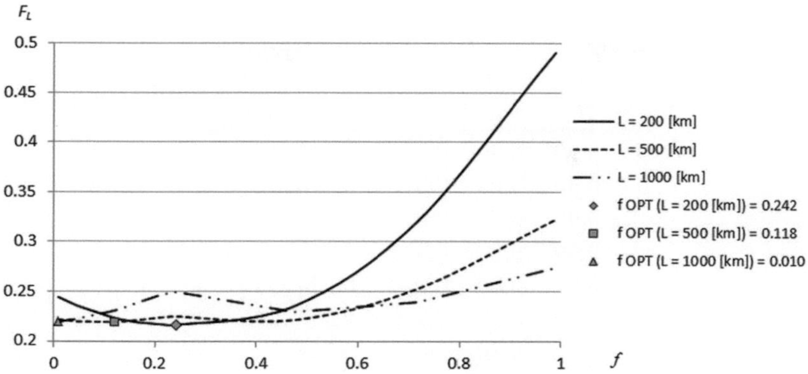

Fig. 3.13 Logistic cost factor (F_L) vs loss factor (f) in case of $\chi = 0.5$, $\psi = 0.9$, $SL = 0.95$, $cv = 0.1$, $c_R = c_N$, and $p = 0.5$ for different transport distances (L)

Table 3.36 SOQ model results in case of $L = 200$ [km] and $c_R = c_N$

$L = 200$ [km]						
					SOQ_N	SS
cv	SL	χ_{OPT}	f_{OPT}	F_L	[unit]	[unit]
0.0	0.90	0.80	0.475	0.208	20	–
0.1	0.90	0.80	0.718	0.210	19	1
	0.95	0.80	0.475	0.210	20	1
	0.99	0.80	0.475	0.210	20	2
0.5	0.90	0.80	0.718	0.213	19	2
	0.95	0.80	0.718	0.211	19	3
	0.99	0.80	0.718	0.210	19	4

Table 3.37 SOQ model results in case of $L = 500$ [km] and $c_R = c_N$

$L = 500$ [km]						
					SOQ_N	SS
cv	SL	χ_{OPT}	f_{OPT}	F_L	[unit]	[unit]
0.0	0.90	0.80	0.162	0.210	20	–
0.1	0.90	0.80	0.010	0.213	22	2
	0.95	0.80	0.010	0.212	19	2
	0.99	0.80	0.010	0.211	19	2
0.5	0.90	0.80	0.989	0.218	20	2
	0.95	0.80	0.718	0.215	23	2
	0.99	0.80	0.010	0.213	19	10

Table 3.38 SOQ model results in case of L = 1000 [km] and $c_R = c_N$

L = 1000 [km]						
					SOQ_N	SS
cv	SL	χ_{OPT}	f_{OPT}	F_L	[unit]	[unit]
0.0	0.90	0.80	0.010	0.211	19	–
0.1	0.90	0.80	0.010	0.214	22	2
	0.95	0.80	0.010	0.212	20	2
	0.99	0.80	0.010	0.211	20	3
0.5	0.90	0.80	0.010	0.224	29	8
	0.95	0.80	0.010	0.218	24	10
	0.99	0.80	0.010	0.214	20	14

In all the cases considered, the model identifies as optimal the repair policy characterized by the highest repair rate value considered ($\chi = \chi_{max}$). As an example, trends of F_L vs. χ are depicted in Fig. 3.14 in case of $\psi = 0.9$, $SL = 0.95$, and $cv = 0.1$ for different transport distances (L). As far as concern the transport means selection, the results obtained proved to be strongly affected by the transport distance. In case of short transport distance (L = 200 [km]) the model identifies as optimal choice trucks with different capacity (see Table 3.36). A slower transport means (f = 0.01) minimizes the total logistics cost in case of a long transport distance (L = 1000 [km], see Table 3.38). Demand variability and service level affect the transport means selection in case of a distance of 500 [km] (see Table 3.37).

Total logistics costs (see F_L values in Tables 3.36–3.38) increase with both the increase of the transport distance (because of transport cost) and the demand variability (mainly because of shortage costs). On the contrary, when high service level values are considered, for a given demand variability and transport distance, the total logistics costs decrease. This happens because of the increase in the SS holding costs is of less magnitude than the reduction in the shortage costs obtained.

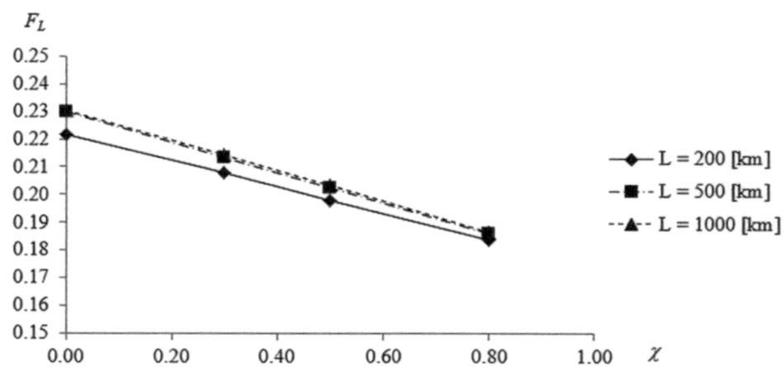

Fig. 3.14 Logistic cost factor (F_L) vs. repair rate (χ) in case of ψ = 0.9, SL = 0.95, cv = 0.1, and $c_R = c_N$ for different transport distances (L)

Finally, the influence of the repair costs on the solution of the logistics problem (3.73) has been also investigated. The problem has been solved for different unit repair costs (c_R) values (see Table 3.34). As an example, trends of F_L versus χ obtained in case of $L = 500$ [km], $cv = 0.1$, and $SL = 0.95$ are depicted in Fig. 3.15.

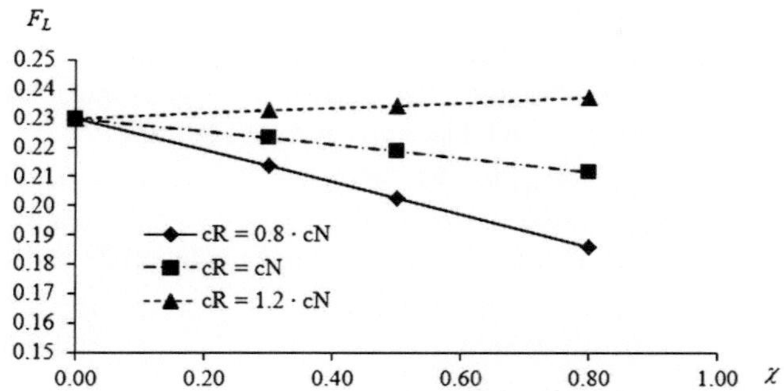

Fig. 3.15 Logistic cost factor (F_L) vs. repair rate (χ) in case of $\psi = 0.9$, $SL = 0.95$, $cv = 0.1$, $L = 500$ [km] for different unit repair costs (c_R)

Similar trends have been obtained for different ψ, SL, cv, and L values. Results showed how the repair policies are cost-effective only in case of a unit repair cost smaller than the unit purchase cost ($c_R \leq c_N$) and high repair success rate (ψ) values.

References

[1] UmweltBundesamt, Probas database [Online]. Available: http://www.probas.umweltbundesamt.de. Accessed 2009, 2012

[2] EcoTransIT, Calculation [Online]. Available: http://www.ecotransit.org/calculation.en.html

[3] H. van Essen, A. Schroten, M. Otten, D. Sutter, C. Schreyer, R. Zandonella, M. Maibach, C. Doll, External costs of transport in Europe, Nov 2011 [Online]. Available: http://www.cedelft.eu/

[4] C. Schreyer, C. Schneider, M. Maibach, W. Rothengatter, C. Doll, D. Schmedding, External costs of transport. Update study, Oct 2004 [Online]. Available: http://habitat.aq.upm.es/boletin/n28/ncost.en.pdf

[5] Istat, Rivalutazioni e documentazione su prezzi, costi e retribuzioni contrattuali (2014) [Online]. Available: http://rivaluta.istat.it/Rivaluta/

[6] ISPRA, Sinanet database [Online]. Available: http://www.sinanet.isprambiente.it/it. Accessed 2012

[7] EEA, Climate for a transport change. Term 2007: indicators tracking transport and environment in the European Union (2008) [Online]. Available: http://www.eea.europa.eu/

[8] UK.DfT, Free flow vehicle speeds in Great Britain 2011 (2012) [Online]. Available: https://www.gov.uk/

[9] M. Gnoni, R. Iavagnilio, G. Mossa, G. Mummolo, Modelling dynamics of a supply chain under uncertainty: a case from the automotive industry". Int. J. Automot Technol Manage **3**(3), 354–367 (2003)

[10] M. Gnoni, R. Iavagnilio, G. Mossa, G. Mummolo, A. di Leva, Production planning of a multi-site manufacturing system by hybrid modelling: a case study from the automotive industry. Int. J. Prod. Econ. **85**(2), 251–262 (2003)

[11] A. Azapagic, S. Perdan, R. Clift, *Sustainable Development in Practice: Case Studies for Engineers and Scientists* (John Wiley & Sons, Chichester, 2004), p. 458

[12] F. de Leeuw, A set of emission indicators for long-range transboundary air pollution. Environ. Sci. Policy **5**(2), 135–145 (2002)

[13] T. Van Harmelen, R. Korenrompa, C. Van Deutekomb, T. Ligthart, S. Van Leeuwenc, R. Van Gijlswijk, The Price of Toxicity. Methodology for the Assessment of Shadow Prices for Human Toxicity, Ecotoxicity and Abiotic Depletion, in *Eco-Efficiency in Industry and Science Quantified Eco-Efficiency: An Introduction with Applications*, ed. by G. Huppers, M. Ishikawa (Springer, Netherlands, 2007), pp. 105–125

[14] J.-P. Rodrigue, C. Comtois, B. Slack, *The Geography of Transport Systems* (Routledge, London, 2009)

[15] F. Harris, How many parts to make at once. Fact. Mag. Manage. **10**(2), 135–136 (1913)

[16] S. Digiesi, G. Mossa, S. Rubino, A sustainable EOQ model for repairable spare parts under uncertain demand. IMA J. Manage. Math. **26**(2), 185–203 (2015)

[17] S. Digiesi, G. Mossa, S. Rubino, *Sustainable Order Quantity of Repairable Spare Parts*. Advanced Maintenance Engineering, IFAC Proceedings Volumes, vol. 2, no. 1 (2013), pp. 181–186

[18] B. Fortunato, G. Mummolo, Technical-economic optimization of a wind power plant by means of a stochastic analytical model. Energy Convers. Manage. **38**(8), 813–827 (1997)

[19] S. Rubino, G. Mossa, S. Digiesi, in *Spare Parts Inventory Reduction: A Multi-Attribute Approach*. Advanced Maintenance Engineering, IFAC Proceedings Volumes, vol. 1, no. 1 (2010), pp. 62–67

Erratum to: New Models for Sustainable Logistics: Internalization of External Costs in Inventory Management

Salvatore Digiesi, Giuseppe Mascolo, Giorgio Mossa and Giovanni Mummolo

DOI 10.1007/978-3-319-19710-4

Erratum to: Salvatore Digiesi et al., *New Models for Sustainable Logistics: Internalization of External Costs in Inventory Management*, SpringerBriefs in Operations Management, DOI 10.1007/978-3-319-19710-4

The copyright holder for this book is:
© The Author(s) 2016
The copyright holder was originally specified as:
© Springer International Publishing Switzerland 2016

The font size of the book title in the spine of the cover has been rectified.

The online version of the book can be found under
http://dx.doi.org/10.1007/978-3-319-19710-4

Salvatore Digiesi et al., *New Models for Sustainable Logistics: Internalization of External Costs in Inventory Management*, SpringerBriefs in Operations Management,
DOI 10.1007/978-3-319-19710-4

Index

S. Digiesi et al., *New Models for Sustainable Logistics*,
SpringerBriefs in Operations Management, DOI 10.1007/978-3-319-19710-4

MIX
Papier aus verantwortungsvollen Quellen
Paper from responsible sources
FSC® C105338

If you have any concerns about our products,
you can contact us on
ProductSafety@springernature.com

In case Publisher is established outside the EU,
the EU authorized representative is:
Springer Nature Customer Service Center GmbH
Europaplatz 3, 69115 Heidelberg, Germany

Printed by Libri Plureos GmbH
in Hamburg, Germany